Key issues in economics and business
Polish up your maths

Key Issues in Economics and Business

Series editors: Alan Griffiths, Keith Pye and Stuart Wall

Polish up your maths

Gordon Bancroft
Mike Fletcher

Longman
London and New York

Longman Group UK Limited,
Longman House, Burnt Mill, Harlow,
Essex CM20 2JE, England
and Associated Companies throughout the world.

Published in the United States of America
by Longman Inc., New York

© Longman Group UK Limited 1987

First published 1987

British Library Cataloguing in Publication Data
Bancroft, Gordon
   Polish up your mathematics. — (Key issues
   in economics and business)
   1. Economics, Mathematical
   I. Title   II. Fletcher, Michael   III. Series
   510'.2433   HB135
   ISBN 0-582-29721-4

Library of Congress Cataloging in Publication Data
Bancroft, Gordon.
   Polish up your mathematics.

   (Key issues in economics and business)
   Includes index.
   1. Economics, Mathematical.   I. Fletcher, Michael.
   II. Title.   III. Series.
   HB135.B36   1987   513'.12   86–15310
   ISBN 0–582–29721–4

Set in Linotron 202 10/11pt Plantin
Produced by Longman Group (FE) Limited
Printed in Hong Kong

Contents

Editors' preface

Each title in this series takes a particular area of economics or business studies and subjects it to rather more scrutiny than is possible in most introductory textbooks. In this case we focus upon the basic mathematical skills which are a pre-requisite for progress in these subject areas. With helpful guidance and much practice students frequently make rapid improvement in mathematical skills which they thought long forgotten or never understood. It is our firm belief that this title will help and encourage many students who approach even the basic mathematical skills with a distinct feeling of unease.

Alan Griffiths
Keith Pye
Stuart Wall

Authors' preface

This book is not aimed at any particular examination but rather to improve, refresh and add to a student's basic mathematical ability. The main purpose of this book is to provide help for students on courses such as Business Studies, Economics or the Social Sciences who are following courses or modules in mathematics and statistics but who have difficulty regaining the mathematical skills they had at school. It can be used both as an introductory text to be studied before the mathematical course or module begins, or throughout the course whenever a particular concept or skill causes difficulty. The design of this book is such that it can be used by a student without any supervision from the lecturer, and the many exercises, all with solutions, make this a very practical book to study from.

Besides this academic use, the book may be of interest to those who wish to improve their mathematics just out of general interest. If you have not studied mathematics for years and fancy a challenge, this book is for you!

The calculator, when used correctly, is a valuable tool for performing calculations, particularly when these calculations are large and time consuming. The use of a calculator is taught throughout the book, and any scientific calculator of moderate cost will prove adequate. Besides the usual $+$, $-$, $\times$ and $\div$ it should have $\sqrt{}$, x^y or y^x and a memory.

Acknowledgements

We gratefully acknowledge the permission of many businesses and government agencies to use relevant material for this book. We also thank the staff and students of North Staffordshire Polytechnic for testing and checking the text, Kate Howard for patiently transforming the scrawl into typescript, and our wives Anne and Sheila for their help and encouragement. Finally we ought to thank the awful British weather for ensuring a speedy completion of the text!

We are grateful to the following for permission to reproduce copyright material:

The Automobile Association for table 1.2; Gower Publishing Co Ltd for figs 7.1 & 7.2 (c) London Business School, Centre for Economic Forecasting 1986; Haymarket Publishing Ltd for tables 6.5 & 6.10; The Controller of Her Majesty's Stationery Office for figs 7.4, 7.5, 7.6, 7.7, 7.8, 7.9 & tables 5.6, 6.6, 6.12 Crown Copyright; International Leisure Group plc for table 6.8; Longman Group UK Ltd for table 6.7; Office of Population Censuses and Surveys for fig 7.3 & tables 9.1, 9.5 Crown Copyright; The Post Office for table 2.3.

We have been unable to trace the copyright holder in table 6.4 and would appreciate any information that would enable us to do.

Part one

Numbers

Unit 1
Arithmetic operations

In all walks of life, whether it be at work, at home or socially, an ability to handle numbers and to perform basic arithmetic is an essential requirement in today's society. Indeed, it is as important to understand basic arithmetic as it is to read and write. In this opening unit of this text the basic operations of arithmetic, namely addition, subtraction, multiplication and division, are described and applied to problems involving whole numbers. Although you may feel confident that this will cause little difficulty, do you know how to compute, for example,

$$15 - 3 \times 2 ?$$

The correct answer is 9 but did you get the answer 24? Many of you will. If, perhaps, brackets had been used it would have made the order of operations more clear. In this case

$$15 - (3 \times 2) = 15 - 6 = 9.$$

The answer 24 is, of course, the correct answer to the problem

$$(15 - 3) \times 2$$

which has the same numbers and the same operation signs but the order of the operations are different. Therefore it is clearly sensible to use brackets to emphasize the order of the arithmetic operations and hence remove any ambiguity that may exist. An essential rule in mathematics, not just arithmetic, is that expressions within brackets are evaluated first. So

$$(10 \times 8) - (15 \times 5) = 80 - 75 = 5.$$

However, this does not help us with the initial problem, $15 - 3 \times 2$, which contains no brackets. How would your calculator cope with this problem?

Take another example

$8 + 4 \times 3$

and key into your calculator

$\boxed{8}$ $\boxed{+}$ $\boxed{4}$ $\boxed{\times}$ $\boxed{3}$ $\boxed{=}$.

Most good calculators will give 20 for the answer as it performs the multiplication operation before the addition. This is standard practice in mathematics and you should always follow this pattern, so

$8 + 4 \times 3 = 8 + 12 = 20.$

Unfortunately some cheaper calculators do work out the expression from left to right giving the answer of 36 in this example. Therefore it is a sensible idea to become familiar with the way that your calculator performs this process. From now on perform the multiplication (or division) operation before addition (or subtraction). To make this more clear the convention of using brackets to replace the multiplication sign is often used. For example, $10 \times 4 - 5 \times 7$ is rewritten in the form $10(4) - 5(7)$ to give 5 as the answer. This convention of using brackets instead of the multiplication sign will be used throughout this text. Besides making the order of operation more clear it also avoids the confusion between the multiplication sign $\times$ and the letter x.

If a problem contains only pluses or minuses (or only multiplication and division) there is not usually any confusion as the expression is evaluated from left to right. For example,

$12 + 3 - 7 + 2 = 15 - 7 + 2 = 8 + 2 = 10.$

Use your calculator to confirm this answer. Consequently the following summary can be used to evaluate a stated arithmetic problem:
1. Use brackets to clarify the correct order of operation.
2. Evaluate expressions within brackets first.
3. Perform multiplication (and division) before addition (and subtraction).
4. Work from left to right if the expression contains only addition and subtraction (or multiplication and division only).
These rules are illustrated now in an example that describes a situation in everyday life.

Colin and Steve go out one evening with their wives, Maureen and Gill, to a local public house, 'The Fox and Hounds'. The price

3

Table 1.1   Fox and Hounds drinks tariff

Drink	Price (pence)
Beer – pint	80
Beer – half pint	41
Lager – pint	85
Lager – half pint	43
Gin	72
Whisky	76
Rum	82
Lime additive	5
Tonic	25
Ginger ale	30
Orange juice	44

list for drinks is displayed by the bar and part of it can be seen in Table 1.1.

Colin ordered the first round of drinks, two pints of lager and lime, one gin and tonic, and one orange juice. The cost of this round can be expressed by

2(85+5) + (72+25) + 44 pence.

You can see that the cost of two lager and limes is given by 2(85+5), showing the usefulness of the brackets notation. Note that the expression $2 \times 85 + 5 = 175$ gives a different answer to the correct expression 2(85+5). The total cost for this round of drinks is, again

2(85+5) + (72+25) + 44.

Evaluating the terms inside the brackets gives

2(90) + 97 + 44.

Performing the multiplication before any addition

180 + 97 + 44.

Now working from left to right

277 + 44 = 321 pence (or £3.21).

Steve bought the second round of drinks; two pints of beer (the lager was too gassy!), one gin and tonic, and a ginger ale. Steve paid for this round with a £5 note, so the amount of change he receives is given by the following arithmetic expression

500 − (2(80) + (72+25) + 30) pence.

Using the rules describing the correct order of operation, Steve's actual change is

$$500 - (2(80) + 97 + 30)$$
$$= 500 - (160 + 97 + 30)$$
$$= 500 - (287)$$
$$= 213 \text{ pence (or £2.13).}$$

Example

A quarterly telephone bill includes both a standing charge of £9.50 plus a further charge of 5 pence per metered unit. Give an arithmetic expression for the total bill if 250 meter units are used. Evaluate the expression.

Solution

Total bill (in pence) = 950 + 250(5)
$$= 950 + 1250 = 2200$$

The telephone bill will be for £22.

Arithmetic operations are further complicated if some of the numbers are negative. A daily weather report might give a maximum daytime temperature of 12 °C and might also give a minimum temperature on that day of −5 °C. The difference between two numbers is computed by subtracting the smaller number from the larger one, so the difference between these two quoted temperatures is given by

$$12 - (-5).$$

The answer is 17 °C, which can be seen more clearly by observing *Fig. 1.1*.

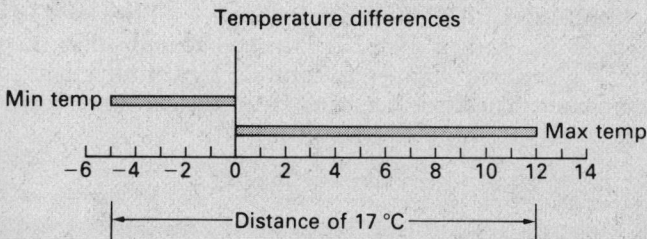

Fig. 1.1

The implementation of a bank account is one situation which requires the application of both positive and negative numbers. Here the value of a credit can be treated as a positive number whereas the amount of a cheque or withdrawal is the negative number.

Example

On 1 June Mrs Price had a balance of £135 in her current account.
(a) During June she paid in amounts of £43 and £52 and wrote cheques to the value of £37, £66, £47, £52 and £62.
What is the balance of her account at the end of June?
(b) During July of the same year there was one credit to the value of £120, but she wrote three cheques, £30 each.
Determine the balance of the account at the end of July.
(c) How much must Mrs Price pay into her account at the end of July to pay off her overdraft?

Solution
(a) As the value of cheques are to be considered to be negative numbers, the balance at the end of June is given by the expression

$$135 + (43) + (52) + (-37) + (-66) + (-47) + (-56) + (-62).$$

This can be evaluated directly on a calculator using the key sequence

| 135 | + | 43 | + | 52 | + | 37 | +/− | + | 66 | +/− | + | 47 |
| + | 56 | +/− | + | 62 | +/− | = |

where the +/− button changes the sign of the number displayed and is therefore useful when inputting negative numbers. The answer displayed is −38, indicating that Mrs Price is £38 overdrawn at the end of June. It is useful to note that the addition of a negative number is equivalent to subtracting a positive number. Therefore her bank balance could have been calculated using the arithmetic expression

$$135 + (43+52) - (37+66+47+56+62)$$
$$= 135 + 95 - 268 = -38.$$

(b) As the bank balance at the beginning of July was −£38 the balance at the end of the month is

$$(-38) + (120) + 3(-30)$$
$$= -38 + 120 - 3(30) = -38 + 120 - 90 = -8.$$

Now the account is overdrawn only by £8.

(c) In order to pay off her overdraft she must add

$$0 - (-8) = 8,$$

that is, £8 to her account.

In this part note that subtracting a negative number is the same as adding a positive number.

$$0 - (-8) = 0 + 8 = 8.$$

The above example illustrates the following rules when adding or subtracting negative numbers:

1. Addition of a negative number is the same as subtraction of the equivalent positive number;
2. Subtraction of a negative number is the same as addition of the equivalent positive number.

What happens when we multiply using negative numbers? Use your calculator to evaluate the expressions in the next example taking care to write down the sign (positive or negative) of your answer. Remember to input the negative number by using the $\boxed{+/-}$ key on your calculator.

Example

(a) $7(-3)$ (b) $(-4)(-5)$ (c) $(-12)(4)$

Solution

(a) The key sequence $\boxed{7}$ $\boxed{\times}$ $\boxed{3}$ $\boxed{+/-}$ $\boxed{=}$ yields an answer of -21.

(b) The key sequence $\boxed{4}$ $\boxed{+/-}$ $\boxed{\times}$ $\boxed{5}$ $\boxed{+/-}$ $\boxed{=}$ gives an answer of $+20$.

(c) The key sequence $\boxed{12}$ $\boxed{+/-}$ $\boxed{\times}$ $\boxed{4}$ $\boxed{=}$ gives -48.

The solution to this example illustrates the following rules concerning multiplication using negative numbers:

1. (positive number) (negative number) = negative number;
2. (negative number) (negative number) = positive number;
3. (negative number) (positive number) = negative number.

These rules always apply for *both* multiplication and division. Perhaps the second of these rules is difficult to comprehend, but in speech a double negative statement, such as *not unusual*, may imply a positive statement, such as *usual*. Now answer the next example without using your calculator.

Example

Evaluate

(a) 8(−4)
(b) (−5) (−2)
(c) (−8) ÷ (−4)
(d) (56) ÷ (−7)

Solution

(a) As 8(4) = 32 and multiplication here is between a positive number and a negative number, we have

$$8(-4) = -8(4) = -32$$

(b) Multiplication of two negative numbers gives a positive answer.

$$(-5)(-2) = 5(2) = 10$$

(c) Division of two negative numbers also gives a positive answer.

$$(-8) \div (-4) = 8 \div 4 = \frac{8}{4} = \frac{2}{1} = 2$$

(d) Here

$$(56) \div (-7) = -\frac{56}{7} = -\frac{8}{1} = -8$$

A further useful skill when dealing with numbers is that of rounding a number to the degree of accuracy appropriate to the situation in hand. Often this has the additional advantage of making the general size of the number more comprehensible. For example, you may have a large win on the football pools which amounts to £251,632.78p. (If you have won this much I cannot think why you are reading this book!) It is quite likely that this figure will be reported in the national press as £250,000. From the reader's point of view this latter figure gives a good idea of the size of the win; it provides a good *approximation* to the actual win. This type of approximation is often carried out with large numbers, or with numbers that have a lot of digits, 372.642934 say. In the pools example the win has been rounded to the nearest ten thousand, or correct to two significant figures (as there are two digits to the left of the zeros).

There are many situations where it is advantageous to estimate the value of a quantity. For example, you may wish to know approximately how long it would take to journey from Stoke-on-Trent to London. Using the AA recommended route the distance by road is 156 miles but it is unlikely that you are travelling from the centre of Stoke-on-Trent to the centre of London so it may be sensible to round the distance to the nearest ten miles. As 156 miles is nearer to 160 miles than 150 miles, the distance between the two cities is 160 miles to the nearest ten miles.

Example

Table 1.2 shows the distance, in miles, between major cities within Great Britain. The distances are given to the nearest mile and are measured along the normal AA recommended route.

Table 1.2 Mileage chart for certain major cities

London				
405	Edinburgh			
155	337	Cardiff		
210	222	200	Liverpool	
170	446	119	250	Exeter

Source: Extract from Mileage chart, *AA Handbook*, 1986–87

(a) Determine the distance between the following pairs of cities rounded to the nearest 10 miles.
   (i) Exeter and Cardiff
   (ii) Edinburgh and Liverpool
   (iii) London and Cardiff
(b) What is the distance between London and Edinburgh to the nearest 100 miles?
(c) What is the distance between London and Liverpool rounded to one significant figure?

Solution
(a) (i) The distance between Exeter and Cardiff is 119 miles which is nearer to 120 miles than 110 miles, so the answer is 120 miles.
   (ii) 222 miles is closer to 220 miles than 230 miles, so the answer is 220 miles.

(iii) The distance between London and Cardiff is 155 miles, which is exactly halfway between 150 miles and 160 miles. Convention states that if this happens then the number should be rounded up rather than down. This gives an answer of 160 miles.

(b) 405 miles is nearer to 400 miles than 500 miles, so choose 400 miles.

(c) The numbers 200 and 300 both have one significant figure (one number to the left of the zeros) and the distance 210 miles is nearer to 200 than 300. The answer is 200.

Exercises

1. Evaluate the following expressions
   (a) $7 + 5(2)$ (b) $3 + 7(4)$ (c) $5(3) + 6$

2. A rock climber started climbing 50 metres below a ledge and ended up 35 metres above the ledge. Express the distance climbed arithmetically and evaluate its answer.

3. Without using a calculator evaluate
   (a) $((-3) + 7) \div 2$ (b) $((-2) - (-4)) 3$ (c) $((-5) + (-3)) \div (-4)$

4. The attendance at a football match was 44,821.
   (a) Round the attendance to the nearest one hundred.
   (b) Round the attendance to the nearest one thousand.

5. Use your calculator to compute
   (a) $350 + 5(25)$ (b) $80(4) + 10$ (c) $24(5) + 18(7)$

6. During the first quarter of 1985 there were 50,963 marriages in England and Wales. What is this value to the nearest thousand?

7. Use your calculator to compute
   (a) $(-40) + (-5)16$ (b) $((-25) + (-365)) \div (-13)$

8. The deepest place in the Pacific Ocean is 36,000 feet below sea-level in the Marianas Trench near Guam. The highest mountain in the Himalayas is Mount Everest at 29,000 feet above sea-level. What is the difference in their 'heights'.

9. Mrs Prince's current account was £52 overdrawn. After an amount was paid into this account the balance was £36 in credit. How much was paid into the account?

10. Evaluate
    $83(36) \div (6 + 4(3))$

11. Your monthly gross salary is £645 and you pay £124 in taxes and £43 in national insurance each month. Calculate your take-home pay.

12. A family bought a dining room table, priced at £323, four dining chairs, priced at £62 each, and two carved chairs, £83 each.
    (a) Write a mathematical expression that determines the total value of dining room furniture.
    (b) Determine the total expenditure.
    (c) Round your answer to (b) to the nearest £10.

Unit 2
Fractions and decimals

In Unit 1 attention was restricted to arithmetic operations involving whole numbers. However a glance at any newspaper may reveal, say, that yesterday's rainfall was 0.65 mm, or that a particular article of clothing is 'one quarter off', or that the exchange rate is 1.355 dollars to the pound. All of these pieces of information contain numbers that are part of a whole. It is important to understand this concept of fractions, whether it be common fractions or decimal fractions.

A *common fraction* is a number written like

$$\frac{3}{5},$$

where the number above the line is called the *numerator* and the number below the line is called the *denominator*. The fraction $\frac{3}{5}$ can be envisaged by splitting the whole into five equal parts and taking three of these parts – see Fig. 2.1.

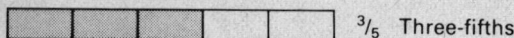

 $^3/_5$ Three-fifths

Fig. 2.1

By considering similar diagrams (see Fig. 2.2) it can be seen that the fractions $\frac{2}{4}$ and $\frac{1}{2}$ are the same.

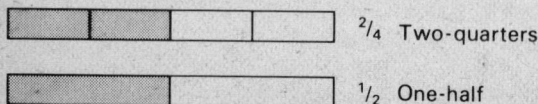

 $^2/_4$ Two-quarters

 $^1/_2$ One-half

Fig. 2.2

We conclude that

$$\frac{2}{4} = \frac{1}{2}.$$

In fact there are many different ways of writing one-half:

$$\frac{1}{2} = \frac{2}{4} = \frac{3}{6} = \frac{4}{8} = \frac{5}{10} = \frac{8}{16}.$$

These are all equal or equivalent fractions and can be formed by multiplying both the numerator and denominator of the fraction by the same constant. It is usually best to reduce a fraction to its simplest form. To do this, divide both the numerator and denominator by the same whole number. When no number other than one will divide into both the numbers, the fraction is said to be in its *lowest form*.

Example

Convert the following fractions to their lowest terms.

(a) $\frac{3}{12}$ (b) $\frac{16}{28}$ (c) $\frac{5}{8}$

Solution

(a) $\frac{3}{12} = \frac{1}{4}$ Both numbers can be divided by 3

(b) $\frac{16}{28} = \frac{4}{7}$ Both number can be divided by 4

(c) $\frac{5}{8}$ is in its lowest form. Although the denominator can be divided by 2 and 4 the numerator can not. Hence it is in its lowest form.

In order to add or subtract two or more fractions they must have the same denominator. Therefore the first step when carrying out this operation is to find the common denominator of all the fractions.

Example

Evaluate

(a) $\frac{1}{5} + \frac{3}{5}$ (b) $\frac{3}{4} + \frac{5}{6}$ (c) $\frac{1}{2} + \frac{1}{3} - \frac{1}{6}$

13

Solution

(a) $\frac{1}{5}$ and $\frac{3}{5}$ both have a denominator equal to 5. Therefore these fractions can be added immediately.

$$\frac{1}{5} + \frac{3}{5} = \frac{1+3}{5} = \frac{4}{5}.$$

(b) $\frac{3}{4}$ and $\frac{5}{6}$ have denominators equal to 4 and 6 respectively. The smallest number divisible by both 4 and 6 is 12. The two fractions need to be rewritten so that they are equivalent fractions with denominator equal to 12.

$$\frac{3}{4} + \frac{5}{6} = \frac{9}{12} + \frac{10}{12} = \frac{9+10}{12} = \frac{19}{12}.$$

The answer $\frac{19}{12}$ is an *improper fraction* (as the numerator is greater than the denominator) and can be written as the mixed number, $1\frac{7}{12}$. A *mixed number* is a combination of a whole number and a fraction.

(c) The smallest number divisible by 2, 3 and 6 is 6 so the fractions are converted to equivalent fractions with denominator equal to 6.

$$\frac{1}{2} + \frac{1}{3} - \frac{1}{6} = \frac{3}{6} + \frac{2}{6} - \frac{1}{6} = \frac{3+2-1}{6} = \frac{4}{6}$$

In its lowest form the answer is $\frac{2}{3}$.

Example

Sixty people were interviewed about their current political opinions. Each of the respondents was asked to name the political party they would vote for if there was a general election on the following day. The results are shown in Table 2.1.

(a) What fraction of the respondents would vote for the two traditional British political parties, Conservative and Labour?

(b) What fraction of the respondents would vote for the three leading political parties?

(c) Which of the parties has the greatest support in this survey?

(d) Determine the fraction of those questioned who intend to vote.

Table 2.1 Survey on political opinion

Political party selected	Fraction of respondents
Labour	$\frac{1}{3}$
Alliance	$\frac{3}{10}$
Conservative	$\frac{4}{15}$
Ecology	$\frac{1}{20}$
Other parties	$\frac{1}{30}$
Will not vote	$\frac{1}{60}$

Solution

(a) $\dfrac{1}{3} + \dfrac{4}{15} = \dfrac{5}{15} + \dfrac{4}{15} = \dfrac{5+4}{15} = \dfrac{9}{15} = \dfrac{3}{5}$

(b) $\dfrac{1}{3} + \dfrac{3}{10} + \dfrac{4}{15} = \dfrac{10}{30} + \dfrac{9}{30} + \dfrac{8}{30} = \dfrac{10+9+8}{30} = \dfrac{27}{30} = \dfrac{9}{10}$

(c) Fractions can only really be compared if their denominators are the same.

$$\text{Labour} \quad \frac{1}{3} = \frac{20}{60}$$

$$\text{Alliance} \quad \frac{3}{10} = \frac{18}{60}$$

$$\text{Conservative} \quad \frac{4}{15} = \frac{16}{60}$$

$$\text{Ecology} \quad \frac{1}{20} = \frac{3}{60}$$

The largest of the numerators is 20, so the Labour party has the greatest support in this survey.

(d) $1 - \dfrac{1}{60} = \dfrac{60}{60} - \dfrac{1}{60} = \dfrac{60-1}{60} = \dfrac{59}{60}.$

To multiply fractions there is *no* need to rewrite the fractions so that they have the same denominator. All you need to do is multiply the numerators together and multiply the denominators together. The resulting fraction is then reduced to its lowest form:

$$\frac{8}{9}\left(\frac{3}{16}\right) = \frac{8(3)}{9(16)} = \frac{24}{144} = \frac{1}{6}.$$

In order to simplify this operation you can divide across the fractions using a common number, a process called *cancellation*. In the above example both 8 and 16 can be divided by 8 and both 3 and 9 can be divided by 3, giving

$$\frac{8}{9}\left(\frac{3}{16}\right) = \frac{\overset{1}{\cancel{8}}(\overset{1}{\cancel{3}})}{\underset{3}{\cancel{9}}(\underset{2}{\cancel{16}})} = \frac{1(1)}{3(2)} = \frac{1}{6}.$$

Even if one of the numbers is a whole number the procedure remains the same, as the whole number, can be expressed as a fraction by giving it a denominator of 1.

$$\frac{2}{3}(12) = \frac{2}{3}\left(\frac{12}{1}\right) = \frac{2(\overset{4}{\cancel{12}})}{\underset{1}{\cancel{3}}(1)} = \frac{8}{1} = 8.$$

The word 'of' has the same meaning as 'multiplied by' in arithmetic. So the above calculation is equivalent to finding two-thirds of 12.

To divide one fraction by another, the second fraction (the *divisor*) is turned 'upside down' (the numerator becomes the denominator and the denominator becomes the numerator) and the two fractions are then multiplied together. Although this sounds a little complicated, it isn't! For example

$$\frac{4}{15} \div \frac{2}{5} = \frac{4}{15}\left(\frac{5}{2}\right) = \frac{\overset{2}{\cancel{4}}(\overset{1}{\cancel{5}})}{\underset{3}{\cancel{15}}(\underset{1}{\cancel{2}})} = \frac{2(1)}{3(1)} = \frac{2}{3}$$

In this example $\frac{2}{5}$ is turned 'upside down' to give $\frac{5}{2}$ which is then multiplied by $\frac{4}{15}$, giving $\frac{2}{3}$ as the answer. In the same way

$$6 \div \frac{3}{5} = \frac{6}{1}\left(\frac{5}{3}\right) = \frac{\overset{2}{\cancel{6}}(5)}{1(\underset{1}{\cancel{3}})} = \frac{2(5)}{1(1)} = \frac{10}{1} = 10$$

Example

In his will John Brain left half of his personal estate to his widow, Gill Brain, and the remainder was split equally between his three children. If the value of his personal estate was £9600 determine how much was received by each member of his family.

Solution

The widow received half of £9600, which is

$$\frac{1}{2}\left(\frac{9600}{1}\right) = £4800.$$

For each child the fraction of the total personal estate is $\frac{1}{3}$ of $\frac{1}{2}$, or

$$\frac{1}{3}\left(\frac{1}{2}\right) = \frac{1}{6}.$$

Each child therefore receives

$$\frac{1}{6}(\cancel{9600}) = £1600.$$

Example

Brian Jackson wishes to buy a house and needs to take out a mortgage to help purchase the home. A large bank will allow Mr Jackson to borrow $2\frac{3}{4}$ times his annual salary, whereas his local building society will only lend him $2\frac{1}{2}$ times his salary. If Mr Jackson's annual salary is £8400 determine the size of loan that each institution will give him.

Solution

The mixed number, $2\frac{3}{4}$, is equal to the improper fraction $\frac{11}{4}$. (The denominator of the improper fraction is the same as the denominator of the fractional part of the whole number, and the numerator is obtained from $2(4) + 3 = 11$.)
From the bank Mr Jackson can borrow

$$2\frac{3}{4}(8400) = \frac{11}{4}(\cancel{8400}) = 11(2100) = £23,100.$$

From the building society he can borrow

$$2\frac{1}{2}(8400) = \frac{5}{2}(\cancel{8400}) = 5(4200) = £21,000.$$

Decimals are fractions expressed in tenths, hundredths, thousandths, etc., and as normal counting takes place in tens, decimals are easier to work with than common fractions. To identify where the whole number ends and the fractional part begins, a decimal point is used, thus 4.8 is equivalent to $4\frac{8}{10}$ or 0.27 is equivalent to $\frac{27}{100}$. When using a calculator decimal numbers are more useful than common fractions; for example, the answer to the calculation

$$\frac{11(3)}{8}$$

can be obtained by using the key sequence

| 11 | × | 3 | ÷ | 8 | = |.

The answer displayed is 4.125, a decimal number. Although many good calculators do have the facility to handle common fractions it is easier to use decimals. The answer, 4.125, to the above problem is equivalent to the common fraction

$$4.125 = 4 + \frac{1}{10} + \frac{2}{100} + \frac{5}{1000}$$
$$= 4\frac{125}{1000}$$
$$= 4\frac{1}{8}$$

in its lowest form.

Since the 1960s the United Kingdom has gradually undergone a process of 'decimalization'. This process has resulted in a much greater use of decimal numbers. It is quite common now to meet statements like
1. the weight of a parcel is 1.35 kg;
2. interest rates are 11.875%;
3. a petrol tank holds 52.8 litres;
whereas in previous years equivalent statements would have used:
(1) 2 lb 13 ounces;   (2) $11\frac{7}{8}$%; and   (3) 11 gallons 5 pints.

As well as being more commonly used nowadays, decimal fractions have the advantage over common fractions that arithmetic is somewhat easier. For example, addition (and subtraction) of decimals is exactly the same as the addition (and subtraction) of whole numbers if a calculator is being used and there should be no problems. Try it!

$$8.32 + 16.1 + 0.2374 = 24.6574$$

If this calculation is carried out by hand it is important to keep the decimal points in a vertical line.

```
  8.32
 16.1
  0.2374
 24.6574
```

Again multiplication of (or division between) two decimal numbers is similar to that between whole numbers, but care must be taken to ensure that the decimal point is in the correct position.

The answer to the problem

$$8.23 \times 6.1 = 50.203$$

has three digits to the right of the decimal point which is the same as the total number of digits to the right of the decimal point in the two numbers 8.23 and 6.1. However, be careful as some of the digits may be zeros. To avoid making very large errors it is safer to estimate the answer when multiplying or dividing decimals. For example, $8.91 \div 7.26$ is just a little more than 1 so if you get the answer of 12.27 you would know this to be wrong.

Example

Evaluate

(a) 27.4(6.3) (b) 100(7.6547) (c) $15.7 \div 4.15$

Solution

(a) The answer will be above 25(6) = 150, but below 30(7) = 210.
27.4(6.3) = 172.62

(b) 100(7.6547) = 765.47
It can be seen that multiplication of a decimal number by 100 simply moves the decimal point two places to the right. Similarly, the decimal point is moved one place to the right if the number is multiplied by 10, or it is moved one place to the left if it is divided by 10.

(c) The answer will be a little less than 4. Using a calculator

$$15.7 \div 4.15 = 3.78313253$$

The answer displayed has 8 digits to the right of the decimal, which is typical of division involving decimal numbers. This degree of accuracy is not usually required and so the answer should be rounded to the degree of accuracy appropriate to the situation. For example,

$$15.7 \div 4.15 = 3.783$$

is correct to four significant figures (or, equivalently, correct to three decimal places).

Example

An electricity bill is made up of a standing charge plus a further charge per unit used. If the standing charge is £8.50 and the cost per unit is 5.32 pence, determine the total cost of the bill if

(a) 150 units,
(b) 450 units,
are used.

Solution
(a) The cost if 150 units are used is, in pence,

$$850 + 150(5.32) = 1648 \text{ (or £16.48)}$$

> *Beware!* It is essential that the same units are used in the calculation. Do not mix pounds with pence.

(b) The cost if 450 units are used is

$$850 + 450(5.32) = 850 + 2394 = 3244 \text{ pence (or £32.44)}$$

Example

A girl school-leaver when looking for employment in a Job Centre notices three situations that appeal to her. The jobs and associated pay are shown in Table 2.2. In order to compare the pay determine the weekly pay of each of the jobs.

Table 2.2 Pay description of three jobs

Job	Pay
Trainee nurse	£316.84 per month
Secretary	£4085.00 per year
Sales assistant	£2.25 per hour (40 hour week)

Solution
Trainee nurse: There are 12 months in a year so the annual pay is

$$316.84 \times 12 = £3802.08.$$

The weekly pay is equal to the annual salary divided by 52, giving

$$3802.08 \div 52 = £73.12,$$

rounded to the nearest penny.
Secretary: The weekly pay is

$$4085 \div 52 = £78.56.$$

Sales assistant: The weekly pay is

$2.25 \times 40 = £90.00$

In terms of weekly pay the sales job is the most preferable.

Exercises

1. The Olympic 800 metres for men was won with a time of 103.0 seconds in 1984 and with a time of 109.2 seconds in 1948. Calculate the difference in the two winning times.

2. How many pieces of wood each $5\frac{1}{2}$ centimetres long can be cut from a strip $123\frac{3}{4}$ centimetres long?

3. Evaluate the following expressions

   (a) $\frac{3}{4} + \frac{5}{8}$ (b) $\frac{5}{9} - \frac{7}{18}$ (c) $\frac{4}{7}\left(\frac{3}{8}\right)$

4. On Monday John worked $6\frac{3}{4}$ hours, on Tuesday he worked $7\frac{1}{2}$ hours, on Wednesday $8\frac{5}{6}$ hours, on Thursday $7\frac{1}{4}$ hours, and on Friday, $5\frac{2}{3}$ hours. What total time did he work? Determine his pay for this week if he is paid at £3.22 per hour.

5. Evaluate the following expressions using decimals
   (a) $9.421 - 5.3264$ (b) $5.71 \div 21.5$ (c) $4.29(7.31 - 4.216)$

6. The circumference of a circle is given by $2\pi r$, where r is the radius of the circle and π is often given as $\frac{22}{7}$. Determine the circumference of the circle with radius $10\frac{1}{2}$ centimetres. Compare the answer with the value of π given by your calculator.

7. A colour television is advertised at £384.95. An alternative method of purchasing the television is to make 24 payments of £17.19. How much extra do you pay by purchasing it on the instalment plan?

8. The national rate for the cost of sending parcels within the United Kingdom is given in Table 2.3. A company wishes to send 8 parcels.

Table 2.3 Inland postal rates

Weight not over	Rate (£)	Weight not over	Rate (£)
1 kg	1.41	7 kg	2.95
2 kg	1.82	8 kg	3.10
3 kg	2.23	9 kg	3.30
4 kg	2.44	10 kg	3.45
5 kg	2.65	25 kg	4.40
6 kg	2.80		

Source: Inland Postal Rates (National), as at Nov. 1985, Post Office

The weight of each of these parcels is 0.85 kg, 1.57 kg, 1.90 kg. 2.74 kg. 4.61 kg, 5.75 kg, 6.40 kg, and 9.40 kg. Calculate the total cost of sending these eight parcels.

9. A litre bottle of whisky costs a publican £7.20. A measure of whisky is 25 ml and is sold at a price of 55 pence.
   (a) Determine the number of measures of whisky per bottle.
   (b) Determine the contribution to profit of selling all of these measures.
   (c) What fraction is the cost of the bottle compared with the money received?

10. A person spends $\frac{1}{4}$ of a 24-hour day asleep, $\frac{3}{8}$ at work, $\frac{1}{12}$ travelling, and the remainder is free time. What fraction of the day is free time?

11. A car achieves 42.7 miles per gallon under normal driving conditions. If the petrol tank holds 9.6 gallons, determine the expected travelling distance on one full tank of petrol.

12. An electrician needs copper wire in lengths of $3\frac{1}{8}$ metres. How many such lengths can be cut from a wire of length 250 metres?

Unit 3
Ratios and percentages

It is often of interest in everyday activities to compare two, or sometimes more, quantities; for example, in a work group there may be twelve apprentices together with three qualified technicians. In this situation we can say that the ratio of technicians to apprentices is 12 : 3. However, as in the case of fractions, it is more usual to express this in its lowest terms. Here 12 and 3 are both divisible by 3 so in its lowest terms the ratio 12 : 3 is expressed as 4 : 1, indicating that in this work group there are four apprentices for each qualified technician. A specific feature of a ratio is that it gives no indication of size; it is merely used as a basis for comparison. In the above example a ratio of 4 : 1 is not only appropriate for 12 apprentices to 3 technicians but the same ratio could be used for 20 apprentices to 5 technicians or 40 apprentices to 10 technicians.

Suppose that a prospective student is considering attendance at one of two Colleges of Further Education, Ayford College or Beeton College. The decision as to which of these two colleges he attends is dependent on two ratios, the male-to-female ratio and the student–staff ratio. The relevant information is provided in Table 3.1

At Ayford College the ratio of male students to female students is

1600 : 800,

Table 3.1   Staff and student numbers

	Ayford College	Beeton College
Number of staff	150	200
Number of male students	1600	1800
Number of female students	800	1200

or, in lowest terms

2 : 1.

At Beeton College the ratio of male students to female students is

1800 : 1200,

or   3 : 2.

Thus the ratio of males to females is greater at Ayford College; that is, there are more males per female at this college even though there are more males at Beeton College.

The total number of students at Ayford is 2400, giving a student–staff ratio of

2400 : 150

or 16 : 1.

Similarly the student–staff ratio at Beeton is

3000 : 200,

or 15 : 1.

On the basis of this information we can expect slightly more personal tuition at Beeton as there the number of students per teacher is 15 compared with 16 at Ayford.

Example

An opinion poll on election party preferences was conducted, with replies received from 1000 interviewees. It was found that 480 favoured the Conservatives, 360 favoured the Labour Party, 120 preferred the Alliance, and the remaining 40 were undecided. Determine

(a) the ratio of decided to undecided,
(b) the ratio of Conservatives to Labour,
(c) the ratio of the two main parties to the rest of the population.

Solution

(a) Altogether there are 960 interviewees with specified party preferences compared with 40 who do not specify a preference. The ratio of decided to undecided is 960 : 40, or

24 : 1.

(b) The ratio of Conservatives to Labour is 480 : 360 or, in lowest terms, 4 : 3.

(c) The number of interviewees who support the two main parties is 840, the remaining 160 either prefer the Alliance or are undecided. Hence the required ratio is 840 : 160 or

21 : 4

Example

A chemical analysis on impurities in river water that passed a given location showed that in 100 parts of impurities, 60 were dangerous, 30 were harmless, and 10 were unidentified. What is the ratio of

(a) dangerous to harmless impurities,

(b) identified to unidentified impurities?

If river flow was such that 140 kilograms of impurities flowed past the given point each day, how many kilograms of each type would be present in the water each day?

Solution

(a) The ratio of dangerous to harmless impurities is 60 : 30 or

2 : 1

(b) The ratio of identified to unidentified impurities is 90 : 10 or

9 : 1.

Amount of dangerous impurities is $\frac{60}{100}$ of 140 kg, giving

$$\frac{60}{100}(140) = 84 \text{ kg.}$$

Amount of harmless impurities is

$$\frac{30}{100}(140) = 42 \text{ kg,}$$

leaving 14 kg that are unidentified.

Ratios are most frequently used in situations where the proportions of ingredients of a specified mixture remain the same no matter what the overall size of the total product. Consequently the amounts of the various ingredients of mixtures such as concrete, fertilizer, recipes and cocktails are often described in ratio form. For example, concrete consists of a mixture of cement, sand and gravel in the ratio 2 parts cement to 3 parts sand and 5 parts gravel.

Example

Given that concrete is made up of its three components cement, sand and gravel in the ratio 2 : 3 : 5 determine how much of each component is needed to make 1500 kg of concrete.

Solution

The total number of parts specified by the ratio is $2 + 3 + 5 = 10$.

Thus $\frac{2}{10} = \frac{1}{5}$ of the mixture is cement, $\frac{3}{10}$ is sand, and $\frac{5}{10} = \frac{1}{2}$ is gravel.

Amount of cement $= \frac{1}{5} (1500)$ kg $= 300$ kg.

Amount of sand $= \frac{3}{10} (1500)$ kg $= 450$ kg.

Amount of gravel $= \frac{1}{2} (1500)$ kg $= 750$ kg.

Example

A particular brand of fertilizer contains 2 parts nitrogen to 1 part potash and 4 parts phosphates by weight. If a specific quantity of this fertilizer contains 4 grams of nitrogen, how much potash and phosphates should it contain?

Solution

According to the ratio the weight of potash is half that of nitrogen, and the weight of phosphates is twice that of nitrogen.

Hence the weight of potash $= \frac{1}{2}(4) = 2$ g, and

the weight of phosphates $= 2(4) = 8$ g.

As with ratios, percentages are also used to represent proportions. The word *percentage* comes from the Latin words 'per centum' meaning 'by the hundred'. Basically, percentages can be considered to be fractions with a denominator of 100, where the numerator appears alone with % after it. For example $\frac{1}{2}$ can be written as $\frac{50}{100} = 50\%$. Similarly $\frac{4}{5} = \frac{80}{100} = 80\%$. In order to change a fraction to a percentage multiply by 100, which is equivalent to converting the fraction to decimal form and moving the decimal point two places to the right.

Example

Convert $\dfrac{7}{20}$ to percentage form.

Solution

Either $\dfrac{7}{20}(100) = 35\%$

or $\dfrac{7}{20} = 0.35 = 35\%$.

Conversely, a percentage is converted to a fraction by dividing by 100, or equivalently take the decimal form and move the decimal point two places to the left.

Example

Convert 62.5% to fraction form.

Solution

$$62.5\% = \dfrac{62.5}{100} = \dfrac{5}{8},$$

or $62.5\% = 0.625 = \dfrac{5}{8}$.

Decimals or fractions bigger than one correspond to percentages greater than 100%. For example, $2\frac{1}{5} = 2.2 = 220\%$.

Example

A workforce can be classified into managerial, skilled and unskilled as in Table 3.2.

Table 3.2 Company staff profile

Class	Number of employees
Managerial	50
Skilled	75
Unskilled	125

Express the number of employees in each class as a percentage of the total workforce.

Solution
The total workforce is 50 + 75 + 125 = 250.
The percentages of employees are:

in the managerial class $= \dfrac{50}{250} = \dfrac{1}{5} = 20\%,$

in the skilled class $= \dfrac{75}{250} = \dfrac{3}{10} = 30\%,$ and

in the unskilled class $= \dfrac{125}{250} = \dfrac{1}{2} = 50\%.$

Example

A survey was carried out to investigate employment activities of the population. In all, 600 individuals were questioned, of which 60% were male. The survey concludes that 90% of the males were in full-time regular employment, and 70% of the females were in full-time regular employment. Out of the 600 who were interviewed, calculate
(a) how many of the males worked full-time,
(b) how many of the females worked full-time.

Solution

60% of 600 $= \dfrac{60}{100} (600) = 360.$ Thus 360 males and 240 females were interviewed.

The number of males in full-time regular employment is

90% of 360 $= \dfrac{90}{100} (360) = 324.$

The number of females in full-time regular employment is

70% of 240 $= \dfrac{70}{100} (240) = 168.$

Percentages have many practical applications in everyday life. For example, in the car, double-glazing and insurance industries, sales persons are often paid on the basis of their success at selling, and receive a *commission* which is a percentage of their sales income. A further important application of percentages involves the idea of discount. In order to stimulate sales a retailer may decide to offer his goods at a price less than its normal price. A *discount* is the monetary amount by which the normal price is reduced expressed as a percentage of the normal price. In addition to these important applications, taxation and investments are also normally expressed in percentage form.

Example

How much money does a salesman earn on a £300 sale if his commission is 15%?

Solution
Salesman's commission is 15% of £300

$$= \frac{15}{100} (300) = £45.$$

Example

An umbrella initially costs £8 but then has VAT added at 15%. The retailer then offers a discount of 25%. Determine the selling price of the umbrella. Does the order of computing the tax and discount influence the price of the umbrella?

Solution
The VAT on the umbrella is 15% of £8

$$= \frac{15}{100} (8) = £1.20$$

So the cost of the umbrella including VAT but before discount is

£8 + £1.20 = £9.20.

The reduction in price due to the discount is 25% of £9.20

$$= \frac{25}{100} (9.20) = £2.30.$$

Hence the price of the umbrella is £9.20 − £2.30 = £6.90.
What would happen if the price after discount is found first and then the tax is added? Would the answer be the same?
If the umbrella cost is £8, then the discount would be 25% of £8 $= \frac{25}{100} (8) = £2$, giving a price after discount but before tax is added of £8 − £2 = £6. The tax can then be computed to be 15% of £6 $= \frac{15}{100} (6) = 0.90$. When this is added to £6 it gives a selling

29

price for the umbrella of £6.90, thus agreeing with the previous answer.

Under most conditions, if a problem involves two percentages it does not matter in which order the percentage calculations are performed.

Exercises

1. Put each of the following ratios in their lowest terms
   (a) 3 : 6   (b) 360 : 48   (c) 68 : 17

2. A college has 3750 students and 125 teachers. What is the student–teacher ratio?

3. A father earns £200 per week and his son earns £120 per week. What is the ratio of their earnings?

4. Green paint is mixed from yellow and blue paint according to the ratio 7 : 3. Calculate how much of each colour is needed to make 50 litres of green paint.

5. A fertilizer contains the active ingredients nitrogen, potash, and phosphorus in the ratio 2 : 1 : 1. If the fertilizer contains 5 g of potash, how much nitrogen and phosphorus does it contain?

6. At the last General Election 34,000 people voted in a local constituency. If they voted for the Alliance, Conservative and Labour parties in the ratio 8 : 3 : 6, how many votes did each party achieve?

7. The scale on a road atlas is given by 1 cm : 2 km. If two towns, Exford and Wyemouth, are 2.5 cm apart on the map, what is the actual distance between these towns?
   Two further towns are known to be 17.5 km apart; what would be the distance between the two towns on the road atlas?

8. Convert each of the following to percentage form
   (a) $\frac{1}{4}$ (b) 0.1   (c) 0.167

9. Convert each of the following percentages to decimals
   (a) 54%   (b) 8.3%   (c) 40%

10. Find
    (a) 50% of 84   (b) 35% of 16   (c) 120% of 45
    (d) 3% of 700   (e) 50% of 70% of 180

11. If 18 men from a shift of 300 are absent from work, what percentage are present?

12. How much money does a salesman earn on a £750 sale if his commission is 2%?

13. A television normally sells for £250 but is currently on sale at a discount of 20%. What is its sale price at the moment?

14. A married couple go out for a meal at a restaurant. The initial cost of the meal is £15 but then VAT is added at 15% and a service charge is added at 10%. What is the total price for the meal?

15. A car costs £4000 new. If it depreciates in value by 12% during the first year, determine its value at the end of this year. If it depreciates by a further 12% during the second year, what is its value after two years? What is this final value as a percentage of the initial cost?

Unit 4

Powers and roots

A multiple-choice test is one in which the respondent has a choice of answers for each question. A typical example of such a test is shown in Table 4.1. Try this test; it should not cause too many problems.

What letters have you put in the five boxes on the right-hand side of the test paper? You should have the letters C,B,A,D,D in this order. If you do not get these answers, try the question again or read the appropriate unit of this text.

In question 1 of the multiple-choice test there are four possible answers even though only one is correct, which is also true in question 2. How many possible arrangements of answers are there for these two questions? For example, one possible arrangement is AA (A for question 1 and A for question 2) even though both answers are incorrect, or AB, etc. It should not take long to check that there are 16 possible arrangements:

AA, AB, AC, AD, BA, BB, BC, BD, CA, CB, CC, CD, DA, DB, DC, DD

An alternative method of determining the possible number of arrangements of answers to these two questions is to calculate

$$4(4) = 16$$

since there are 4 ways of selecting an answer to each of the two questions. Similarly, if all five questions on the test are considered the total number of possible answers is

$$4(4)(4)(4)(4)$$

which can be computed on a calculator using the key sequence

$$\boxed{4} \ \boxed{\times} \ \boxed{4} \ \boxed{\times} \ \boxed{4} \ \boxed{\times} \ \boxed{4} \ \boxed{\times} \ \boxed{4} \ \boxed{=}$$

and gives the answer 1024. A shorthand way of writing this is 4^5,

32

Table 4.1 Numeracy multiple-choice test

Arithmetic test

For each question one of the answers is correct. Write A,B,C or D in the box to the right of the question.

Question 1
Evaluate $5 - 3(4) + 5$ ☐
  A: 22   B: 13
  C: -2 D: 18

Question 2
Evaluate $\dfrac{3}{19}\left(\dfrac{2}{9}+\dfrac{5}{6}\right)$ ☐

  A: $\dfrac{7}{19}$ B: $\dfrac{1}{6}$

  C: $\dfrac{1}{3}$ D: $\dfrac{6}{19}$

Question 3
Calculate $2.531\,(4.2) - 3.707$
correct to 2 decimal places ☐
  A: 6.92   B: 1.248
  C: 6.923   D: 6.93

Question 4
Find 15% of £23.60 ☐
  A: £15.73 B: £3.45
  C: £1.57   D: £3.54

Question 5
A camera normally selling for £102.40 is on sale
at a discount of 25%. What is its sale price? ☐
  A: £25.60 B: £78.60
  C: £75.50 D: £76.80

---

where 4 is the *base* and 5 is the *power*, and is called 'four raised to the power 5'. Therefore

$$4^5 = 4(4)(4)(4)(4) = 1024$$

If your calculator possesses an $\boxed{x^y}$ key or $\boxed{y^x}$ key, then 4^5 can be computed directly using the key sequence

$\boxed{4}$ $\boxed{x^y}$ $\boxed{5}$ $\boxed{=}$

33

Similarly,

$$3^4 = 3(3)(3)(3) = 81$$
$$2^6 = 2(2)(2)(2)(2)(2) = 64$$
$$7^2 = 7(7) = 49$$

Note that when the base is ten, the product is easy to find,

$$10^2 = 10(10) = 100$$
$$10^3 = 10(10)(10) = 1000$$
$$10^4 = 10(10)(10)(10) = 10,000$$

because the power is always exactly equal to the number of zeros in the answer. It follows that $10^1 = 10$ and $10^0 = 1$. In general any number to the power of 1 is equal to itself and any number to the power of 0 is equal to 1. Thus

$$2^1 = 2 \qquad\qquad 2^0 = 1$$
$$5^1 = 5 \qquad\qquad 5^0 = 1.$$

Example

Use the $\boxed{x^y}$ key on your calculator to evaluate

(a) 6^3 (b) 8^1 (c) $(-2)^5$
(d) 7^0 (e) $(3^2)(3^3)$ (f) $4^5 \div 4^2$

Solution
(a) $6^3 = 216$
(b) $8^1 = 8$
(c) $(-2)^5 = -32$

Some calculators display an error message when using the $\boxed{x^y}$ key for negative values of x. If this is the case evaluate 2^5 and change the sign of the answer if the power is odd, as is the case here.
(d) $7^0 = 1$
(e) There are two ways of getting the correct answer here.

$$(3^2)(3^3) = (9)(27) = 243.$$

Alternatively

$$(3^2)(3^3) = 3(3)3(3)(3) = 3^5 = 243$$

(f) Again there are two methods of solution.

$$4^5 \div 4^2 = 1024 \div 16 = 64$$

or

$$4^5 \div 4^2 = \frac{\cancel{4}(\cancel{4})(4)(4)(4)}{\cancel{4}(\cancel{4})} = 4^3 = 64$$

The last two parts of this example illustrate an important property of powers, namely, that if the bases are the same, multiplication is achieved by adding the powers.

$$3(3^3) = 3^1(3^3) = 3^{1+3} = 3^4 = 81$$
$$2^3(2^4) = 2^{3+4} = 2^7 = 128.$$

Similarly, provided the bases are the same, division is achieved by subtracting the powers.

$$8^3 \div 8 = 8^{3-1} = 8^2 = 64$$
$$6^5 \div 6^2 = 6^{5-2} = 6^3 = 216.$$

Do not forget that the bases must be the same for this property to apply. You cannot calculate $2^3 \times 3^4$ in this way.
Using the division rule

$$3^3 \div 3^5 = 3^{3-5} = 3^{-2},$$

but what does 3^{-2} mean? Remember $3^3 \div 3^5$ can also be computed from first principles

$$\frac{3^3}{3^5} = \frac{3(3)(3)}{3(3)(3)(3)(3)} = \frac{1}{3^2}$$

This provides a definition of 3^{-2}, namely

$$3^{-2} = \frac{1}{3^2} = \frac{1}{3(3)} = \frac{1}{9}$$

Similarly

$$7^{-3} = \frac{1}{7^3} = \frac{1}{7(7)(7)} = \frac{1}{343},$$

$$5^{-1} = \frac{1}{5^1} = \frac{1}{5}.$$

All of these answers can be obtained directly using the $\boxed{x^y}$ key on a calculator, although the answers are given in decimal form. For example, 5^{-1} is evaluated using the key sequence

$$\boxed{5} \quad \boxed{x^y} \quad \boxed{1} \quad \boxed{+/-} \quad \boxed{=}$$

and gives the answer 0.2.

Polish up your maths

The inverse of raising to a power is finding the *root* of a number. For example,

$$\sqrt{49} = \text{square root of } 49 = 7 \text{ since } 7^2 = 49$$

$$\sqrt[3]{64} = \text{cube root of } 64 = 4 \text{ since } 4^3 = 64$$

$$\sqrt[4]{81} = \text{fourth root of } 81 = 3 \text{ since } 3^4 = 81.$$

Use the $\boxed{x^y}$ key on your calculator to evaluate $49^{\frac{1}{2}}$, $64^{\frac{1}{3}}$, $81^{\frac{1}{4}}$. The answers are, as above, 7, 4, and 3. For example,

$$\boxed{81} \ \boxed{x^y} \ \boxed{0.25} \ \boxed{=}$$

gives 3, but note that the fractional powers $\frac{1}{2}$, $\frac{1}{3}$, $\frac{1}{4}$ must be input as decimals. To overcome this problem use the $\boxed{x^{1/y}}$ key to evaluate roots.

Example

Use your calculator to evaluate the following expressions:
(a) $4^2(4^4)$ (b) $2^3(2^{-2})$ (c) $3^3 \div 3^6$
(d) $\sqrt{30}$ (e) $\sqrt{-16}$ (f) $2^{\frac{3}{2}}(2^{\frac{5}{2}})$
(g) $1.57^{\frac{1}{3}}$ (h) 0.8^{-2}

Solution
(a) $4^2(4^4) = 4^6 = 4096$
(b) $2^3(2^{-2}) = 2^1 = 2$
(c) $3^3 \div 3^6 = 3^{-3} = \dfrac{1}{3^3} = \dfrac{1}{27}$
(d) Each of the following three key sequences will give the answer $\sqrt{30} = 5.477$ to four significant figures.

   (i) $\boxed{30} \ \boxed{\sqrt{}}$
   (ii) $\boxed{30} \ \boxed{x^{1/y}} \ \boxed{2} \ \boxed{=}$
   (iii) $\boxed{30} \ \boxed{x^y} \ \boxed{0.5} \ \boxed{=}$

(e) The square root of a negative number does not exist. (The calculator displays an error message.)
(f) $2^{\frac{3}{2}}(2^{\frac{5}{2}}) = 2^{\frac{3}{2}+\frac{5}{2}} = 2^4 = 16$
   The addition property of powers applies even if the powers are fractions.
(g) Using the key sequence

$$\boxed{1.57}\;\boxed{x^{1/y}}\;\boxed{3}\;\boxed{=}$$

the answer is displayed as 1.162, to four significant figures.
(h) The key sequence

$$\boxed{0.8}\;\boxed{x^y}\;\boxed{2}\;\boxed{+/-}\;\boxed{=}$$

displays the answer 1.5625.

Example

A common application of powers is in the calculation of compound interest. If an amount P is invested at an interest rate of r% per annum, the resulting amount after n years is given by

$$P\left(1 + \frac{r}{100}\right)^n$$

Use this formula to determine the value of the following two investments:
(a) £500 invested at 11% p.a. for three years,
(b) £2000 invested at 9% p.a. for six months.

Solution
(a) $P = 500, r = 11, n = 3$
$$\text{Value} = 500\left(1 + \frac{11}{100}\right)^3 = 500(1.11)^3$$
$$= 500(1.367631)$$
$$= £683.82 \text{ (to the nearest penny)}$$
(b) $P = 2000, r = 9, n = \frac{1}{2}$
$$\text{Value} = 2000\left(1 + \frac{9}{100}\right)^{1/2}$$
$$= 2000(1.09)^{\frac{1}{2}}$$
$$= 2000(1.04403)$$
$$= £2088.06 \text{ (to the nearest penny)}$$

Example

In Great Britain most families require a long-term loan (or mortgage) to purchase their home. Mortgages are usually repaid in monthly instalments each of which includes interest on the unpaid balance and a payment on the principal (the amount borrowed).

Polish up your maths

The size of the annual repayments is given by the formula

$$\text{Annual repayment} = \frac{P(r)}{100} \frac{(1+r/100)^n}{[(1+r/100)^n - 1]}$$

where

P = principal (size of loan),
r = net interest rate (%),
n = term of repayment in years.

The monthly repayment is then calculated by dividing the annual repayment by 12. Determine the monthly repayments for each of the loans listed in Table 4.2.

Table 4.2 Mortgage descriptions

Name	Loan (£)	Term (years)	Net interest rate (%)
Mr Arrowsmith	45 000	20	9.5
Mrs Brown	15 000	30	8
Mr Cheadle	25 000	25	7.5

(Note that the usual interest rate quoted on mortgage loans is the gross interest rate but as such loans are often liable to tax relief the net interest rate may be the one used in repayment calculations.)

Solution
For Mr Arrowsmith,

$P = 45,000, r = 9.5, n = 20$.

As $\left(1 + \frac{r}{100}\right)^n = \left(1 + \frac{9.5}{100}\right)^{20} = 1.095^{20} = 6.141612104$

(It is important to keep as much accuracy as possible in the intermediate steps of a calculation. Indeed, you could store this number in the memory of your calculator to save keying it in again.)

$$\text{Annual repayment} = \frac{45,000(9.5)}{100}\left(\frac{6.141612104}{5.141612104}\right)$$
$$= £5106.45$$

Mr Arrowsmith's monthly repayment $= \frac{5106.45}{12} = £425.54$.

For Mrs Brown, $P = 15,000$, $r = 8$, $n = 30$, giving an annual repayment of

$$\frac{15,000(8)}{100}\left(\frac{1.08^{30}}{1.08^{30}-1}\right)$$

$$= 1200\left(\frac{10.06265689}{9.06265689}\right) = £1332.41$$

Her monthly repayments are $\dfrac{1332.41}{12} = £111.03$.

The annual repayment for Mr Cheadle is

$$\frac{25,000(7.5)}{100}\left(\frac{1.075^{25}}{1.075^{25}-1}\right) = £2242.76$$

His monthly repayments are £186.90

Finally in this unit, properties of numbers written in scientific notation are investigated. A number is said to be written in *scientific notation* if it is expressed as a number between 1 and 10 together with some power of 10. For example, 250 can be written as $2.5(10^2)$ or 0.036 can be written as $3.6(10^{-2})$. The main advantage of this notation is that very large or very small numbers can be written more compactly.

$$486,000 = 4.86(10^5)$$
$$0.000249 = 2.49(10^{-4})$$

Most calculators use this notation when giving answers that are very large or small. Use your calculator to evaluate

$$0.0023(0.06)$$

It is likely that your calculator displays $1.38 -04$, indicating that the answer is 1.38×10^{-4}, the power of 10 moving the decimal point 4 places to the left

$$1.38 \times 10^{-4} = 0.000138$$

Now check the calculation

$$83000(165400) = 1.37282(10^{10})$$
$$= 13,728,200,000$$

Exercises

1. Evaluate
   (a) 8^3
   (b) 9^0
   (c) $(-2)^4$
   (d) 10^4
   (e) 7^1

Polish up your maths

2. Use your calculator to evaluate the following expressions.
   (a) $4(4^2)$ (b) $6^3 \div 6^5$ (c) $7^2(7^{-3})$
   (d) $(-3)^{-2}$ (e) $3\frac{1}{2}(3^2\frac{1}{2})$

3. A multiple-choice test, comprising 10 questions, allows five possible answers for each question. Determine the total number of arrangements of possible answers for this test.

4. Evaluate the following expressions.
   (a) $\sqrt{64}$ (b) $\sqrt{40}$ (c) $\sqrt{-25}$
   (d) $12.6\frac{1}{2}$ (e) $7.98\frac{1}{3}$

5. An insurance company purchased a number of securities valued at £10,000 that guaranteed an interest rate of 8% per annum. If these securities were kept for three years, how much interest will they earn?

6. Evaluate
   (a) 0.01^2 (b) $\sqrt{0.01}$ (c) 0.01^{-1}

7. £750 was invested in a bank account at 9.5% per annum. Determine the value of this investment after 18 months.

8. A family takes out a mortgage over 25 years for £50,000. If the net interest rate is 8.4% calculate their monthly repayments.

9. Express the following numbers in ordinary decimal notation:
   (a) $5.423(10^4)$ (b) $7.29(10^1)$ (c) $8.61(10^{-2})$

10. Subtract $9.324(10^{-2})$ from $4.75(10^{-1})$ and give the answer in scientific notation.

Unit 5
Averages

In daily life it is not unusual to meet phrases similar to
  'The average cooked meal contains 800 calories,'
  'The average lifetime of a light bulb is 850 hours,'
  'The average number of children per family is 2.4.'
But is it clear what is meant by this term 'average'? Broadly speaking, the term is used to replace a collection of measurements by just one value. Indeed, the average of a set of numbers may be thought of as a number somewhere at the centre of the set and typical of the set as a whole. However, the word 'average' can be used to deliberately mislead or confuse as the word does have more than one meaning. In this unit three measures of average are described; they are called the mean, the median, and the mode.

Firstly, the *mean*, or the arithmetic mean as it is sometimes called, is found by adding the set of numbers together and dividing by the number of items involved. For example, the mean of the numbers 9,5,3,5,6 and 8 is

$$\frac{9+5+3+5+6+8}{6} = \frac{36}{6} = 6.$$

Suppose the numbers are 50, 90, 60, 70, 60. These have a mean equal to

$$\frac{50+90+60+70+60}{5} = \frac{330}{5} = 66.$$

To get a feel for this, the above can be interpreted in terms of the following:

Andrew has 50p, Richard has 90p, Claire has 60p, Stephen has 70p, and Elizabeth has 60p. How much would each receive if they put their money into a pool and then took out an equal share of the pool? The amount of the pool would be 50+90+60+70+60 = 330 pence. There are five of them to share it and so the share each would have is 330/5 = 66 pence.

It is usual to represent the mean and its calculation by a mathematical shorthand. In the calculation of the mean there are two operations:
1. Add all the numbers together;
2. Divide by the number of values.

In the above example there are five numbers 50,90,60,70,60. Suppose, more generally, there are n numbers, represented by

$$x_1, x_2, x_3, \ldots, x_n.$$

The shorthand notation for 'add all the x-values together' is Σx (read 'sigma x'). For example,

$$x_1 = 5, x_2 = 12, x_3 = 8, x_4 = 5, x_5 = 7, x_6 = 4, x_7 = 10$$
$$\Sigma x = 51,$$

obtained using the key sequence

$$\boxed{5} \ \boxed{+} \ \boxed{12} \ \boxed{+} \ \boxed{8} \ \boxed{+} \ \boxed{5} \ \boxed{+} \ \boxed{7} \ \boxed{+} \ \boxed{4} \ \boxed{+} \ \boxed{10} \ \boxed{=} .$$

The sum is then divided by the number of items in order to find the mean of x. This is often written $\bar{x}$ (read 'x–bar'). So the mathematical shorthand for the mean of a set of data is

$$\bar{x} = \frac{\Sigma x}{n}.$$

If you come across the word 'average' in any technical publication or announcement it will probably be the mean that is being referred to.

Example

A survey took place to investigate the prices of apples and oranges in a certain town. The investigator went into 15 greengrocer shops on one Saturday and the prices he observed are recorded in Table 5.1.

Table 5.1 Shop prices (in pence)

Shop	A	B	C	D	E	F	G	H	I	J	K	L	M	N	O
Apples (1 kg)	45	50	56	60	44	48	39	52	49	64	49	55	64	40	50
Oranges (each)	12	10	13	15	12	12	8	12	14	15	12	14	9	10	9

Determine
(a) the mean price of 1 kg of apples,
(b) the mean price per orange,
in these fifteen shops.

Solution

(a) Using the formula $\bar{x} = \dfrac{\Sigma x}{n}$ for the apple prices,

$$\Sigma x = 45+50+56+ \ldots +50 = 765$$
and so $\bar{x} = \dfrac{765}{15} = 51.$

The mean price of one kilo of apples in this town is 51p.
(b) For orange prices

$$\Sigma x = 12+10+13+ \ldots +9 = 177$$
$$\bar{x} = \frac{177}{15} = 11.8.$$

The mean price of one orange is 11.8 pence. Do not be put off by the fact that it is impossible for an individual orange to cost 11.8p, it does provide useful information about the price of an orange.

In the above example the prices of the oranges might have been recorded in a *frequency table*, as in Table 5.2.

This table shows, for example, that 5 out of the 15 shops sold their oranges at 12 pence each. From the definition of the man,

$$\bar{x} = \frac{\text{total of orange prices}}{\text{number of oranges}}$$

Table 5.2 Frequency table of orange prices

Price per orange	Number of shops
8	1
9	2
10	2
12	5
13	1
14	2
15	2
Total	15

From the table, the number of oranges is 15. The total of the prices for these oranges is

$$1(8)+2(9)+2(10)+5(12)+1(13)+2(14)+2(15) = 177.$$

The mean $= \dfrac{177}{15} = 11.8$, as before.

The second type of average is the *median*, or middle value, where half of the numbers are less than or equal to the median and half are more than or equal to the median. To find the median of a set of numbers, arrange them in ascending order of size and pick out the one in the middle. For example,

2,4,9,7,5   have median 5,

for arranged in order they are 2,4,5,7,9 and 5 is in the middle. Another example is

1,2,4,5,6,6,8,9,10   which have median 6.

These are already in ascending order and the value of the one in the middle is 6. It does not matter that there is another 6 as well.

In both of these examples there have been an odd number of items. If the number of items is even, the definition given at the beginning has to be modified slightly. Then the values are arranged in ascending order and the two in the middle are picked out. The median is then found by computing the mean of these two numbers. For example,

11,8,6,2,3,12,10,14   have median 9

for arranged in order they are 2,3,6,8,10,11,12,14; the two in the middle are 8 and 10, and these two numbers have mean 9.

Example

Table 5.3 shows the number of children in 30 households.

Table 5.3 Number of children per household

2	1	0	7	4	1	2	0	2	3
3	4	2	1	2	5	1	3	2	4
2	1	0	2	3	3	1	2	3	2

Determine the median number of children per household.

Solution

In ascending order the data is

0,0,0,1,1,1,1,1,1,2,2,2,2,2,2,2,2,2,2,2,3,3,3,3,3,3,4,4,4,5,7

There are 30 numbers, so the median is given by the mean of the 15th and 16th numbers. Counting from the left the 15th number is 2, as is the 16th number.

Median = 2

thus indicating that an average or typical family has 2 children.

If there are a large number of items it may be easier to find the median by alternatively striking out the largest and smallest numbers. Using this last example

```
2 1 0 7 4 1 2 0 2 3
3 4 2 1 2 5 1 3 2 4
2 1 0 2 3 3 1 2 3 2
```

first strike out the largest in the set (7), then strike out the smallest (0), next strike out the second largest (5), then strike out the second smallest (0). Continue in this way, alternatively striking out the largest remaining value and the smallest remaining value until one number (or two numbers if the number of items is even) is left.

The third type of average is the *mode* which is used to indicate the value (or type) that occurs most often. The mode is an 'average' in the sense that it is the most common. It is found by counting the number of times each value occurs and then selecting the value with the highest frequency. For example, the numbers

4,8,12,10,7,5,8,14,20

have a mode equal to 8 because it occurs twice and no other number occurs more than once.

Example

A worker at a post office keeps a record of the postage stamps sold at his branch. In one day 160 postage stamps were sold to its customers. The cost and frequency of each is shown in Table 5.4. Determine the mode.

Solution

The postage stamp most frequently bought is that which costs 17p. This is the mode.

Table 5.4 Type and frequency of postage stamps

Type of postage stamp	Frequency
12p	45
17p	64
18p	10
24p	4
31p	2
50p	7
100p	18

We have seen how the mean, median and mode can be obtained from a set of numbers. All three of these measures can be used to indicate the 'average' of the numbers in their different meanings. Consequently the word 'average' might lead to confusion if the type of average is not indicated. Consider the following example.

Example

The annual income of twenty households in one street is shown in Table 5.5. In each case the annual income is recorded to the nearest one hundred pounds.

Table 5.5 Annual incomes (£)

5 400	6 000	6 000	6 300	6 600
7 200	7 500	9 000	9 000	9 000
9 600	10 200	10 800	12 000	12 000
13 500	14 400	15 000	18 000	22 500

Determine (a) the mean, (b) the median, (c) the mode.

Solution

(a) The total of all annual incomes is
$$\Sigma x = 5400 + 6000 + 6000 + \ldots + 22,500 = 210,000$$
The mean income $= \dfrac{210,000}{20} = £10,500$

(b) The incomes are already in ascending order.
The median income $= \dfrac{9000+9600}{2} = £9300$

(c) The income that occurs most frequently, and hence the mode, is £9000.

These three values are all different, emphasizing the need for clarity when using the term 'average'.

Exercises

1. The 1984 temperature, rainfall, and sunshine figures for England and Wales are available for each month, and are shown in Table 5.6.

Table 5.6 Temperature, rainfall, and sunshine for 1984 in England and Wales

Months in 1984	Mean daily air temp. (°C)	Rainfall (mm)	Mean daily sunshine (hours)
Jan.	3.8	144	2.18
Feb.	3.5	57	2.14
Mar.	4.6	59	2.09
Apr.	8.5	11	7.42
May	10.1	59	5.58
June	14.7	43	7.03
July	16.6	27	7.89
Aug.	17.2	57	6.53
Sept.	13.7	116	3.76
Oct.	11.2	100	3.17
Nov.	8.7	145	1.82
Dec.	5.7	81	1.76

Source: Monthly Digest of Statistics, Nov. 1985

Assuming that each month has equal weight determine
(a) the mean daily air temperature,
(b) the mean rainfall,
(c) the mean daily sunshine
for 1984 as a whole.

2. A footwear shop sells sports shoes in whole sizes only. One summer, 260 sports shoes were sold at this shop. A breakdown of the shoe sizes sold can be seen in Table 5.7.
   What is the modal size of shoe?

3. One cricket season Alan Smith batted on eighteen occasions. His scores were
   9,32,23,0,56,28,4,12,66,84,59,3,0,42,19,8,22,26
   Determine the mean score.

Polish up your maths

Table 5.7 Frequency table of shoe sizes

Shoe size	Frequency
3	8
4	12
5	25
6	43
7	65
8	57
9	32
10	18

4. Brian travelled to work by car. His journey times, in minutes, over a three-week period are shown in Table 5.8.

Table 5.8 Journey times (minutes)

	Mon	Tues	Wed	Thurs	Fri
Week 1	42	38	39	36	47
Week 2	43	41	35	38	44
Week 3	41	36	37	39	42

Determine the median journey time.

5. Eight different tour operators organize a 14-day holiday to a certain hotel in a Spanish holiday resort. The quoted price of each of these holidays is, in pounds,
384,396,354,387,389,399,355,360
Determine the mean price of a tour to this hotel.

6. At eight performances of a play in a week the numbers present were
873,681,752,942,621,826,1036,1092
Determine the median attendance per performance.

7. Petrol prices at twelve different garages in a given town centre were
£1.77 £1.69 £1.72 £1.77 £1.79 £1.80
£1.77 £1.77 £1.73 £1.78 £1.77 £1.76
Determine (a) the mean, (b) the median, (c) the mode.
Which of these averages do you think is the most appropriate?

8. For the data in Table 5.3 calculate the mean and mode and compare with the median.

Part two

Pictures

Unit 6
Tables

Generally the collection of information from a survey or experiment leads to a large mass of data. In their original form, as a large set of numbers, these data appear almost meaningless. They need to be condensed into a more understandable form. The figures in Table 6.1 represent the number of matches inside a sample of 50 match-boxes.

Table 6.1 Number of matches per matchbox

232	251	237	255	229	246	244	249	241	235
250	248	245	246	234	239	250	242	240	246
238	249	247	243	236	250	248	244	247	239
252	247	246	241	244	245	245	243	237	246
254	252	245	243	242	244	246	242	239	244

The first step towards making it more presentable is to convert it to the form of a *grouped frequency table*. This involves dividing the range of the data into classes and counting the number of items which fall into each class. The data here ranges from 229 to 255 and suitable classes are as shown in Table 6.2.

There are obviously many different ways in which the range 229 to 255 can be divided. In the classification given in Table 6.2 there are nine classes, all of equal size. There are no firm rules about how many classes to use but it is usual to have between 5 and 15 classes. Less than 5 would create a loss in information whereas more than 15 would be too many to take in easily and so would rob the frequency table of its main reason for existence. In addition, if the classes are of equal size it gives a clearer view of the distribution of the data.

Having set up the classes the next step is to count the number of

Table 6.2   Suitable classes

Number of matches per box
229–231
232–234
235–237
238–240
241–243
244–246
247–249
250–252
253–255

data items falling into each. The usual way is to employ a tallying procedure. This involves taking each item in turn in the original data and entering a tally-mark against the appropriate class. It is a good idea to record the tally-marks using a 'five-bar gate' method so that they can be easily added up at the end. The number of tally-marks recorded against a class is the class frequency. The above example produces the grouped frequency table shown in Table 6.3.

Table 6.3   Frequency table

Class	Tally-marks	Frequency
229–231	1	1
232–234	11	2
235–237	1111	4
238–240	IIII	5
241–243	IIII 111	8
244–246	IIII IIII IIII	15
247–249	IIII 11	7
250–252	IIII 1	6
253–255	11	2

This example gives a good indication of how to construct a table from statistical data; however, in daily life it is more likely that an individual will have to read various types of table and interpret the information they contain. Many tables are straight forward to understand but some are much more complicated. It is essential that today's citizen can understand a set of tabulated data as so

Table 6.4 BUDGET car rental charges

Vehicle Group	Vehicle type	1,2 Day Excess mileage charge	Daily 1,6 days	Weekly 7 days	Daily 8 days	Excess Hourly charge	Special Weekend rate
A	Ford Fiesta, VX Nova, Nissan Micra		£19.50	£108.00	£15.90	£3.90	£39.00
B	Ford Escort, VX Astra, Metro (Auto), Nissan Cherry		£21.50	£118.00	£17.15	£4.30	£43.00
C	Ford Sierra, VX Cavalier, Nissan Stanza A/Montego, MG Metro		£25.00	£137.50	£20.00	£5.00	£50.00
D	Peugeot 205 GTI Ford Orion Ghia (Auto) VX Cavalier Estate		£33.75	£182.00	£26.40	£6.75	£67.50
E	Ford Sierra 2.0 GL (Auto) VX Cavalier SRi, Escort XR3i	17p	£39.50	£210.00	£30.50	£7.90	£79.00
F	Mercedes 190E (Auto) BMW 320 (Auto)	24p	£59.00	£337.00	£48.50	£11.80	£118.00
G	Mercedes 230E (Auto) BMW 525E (Auto)	26p	£78.75	£460.00	£66.00	£15.75	£157.50
H	Mercedes 280SE (Auto), Jaguar XJ6 4.2 (Auto), Range Rover c/w Tow Bar	28p	£99.50	£600.00	£87.00	£19.90	£199.00
I	Porsche 911SC Coupe	65p	£96.00	£620.00	£90.00	£19.20	£192.00
J	Rolls Royce	80p	£260.00	£1440.00	£228.00	£52.00	£520.00

Notes: 1. VAT at 15% to be added 2. Unlimited mileage groups A to D 3. Groups E to J, 1–2 day, 100 free miles

much information is presented in this form. The rest of this unit looks at some common examples of tabulated information and in each case the information which is relevant is extracted. The tables in these examples display some or all of the features that well-presented tables must have to give clear and unambiguous information. These are
1. a title,
2. clearly stated units,
3. a source, if appropriate,
4. footnotes to help clarification.

Example

A well-known international car rental firm issued information shown in Table 6.4 showing its rent-a-car rates for 1985.

Additionally the petrol consumption of three of the cars are given in Table 6.5.

Table 6.5 Fuel consumption

Car	Miles per gallon
Range Rover	15
Nissan Cherry	30
Escort XR3i	28

Source: What Car? Apr. 1986

Assuming one gallon of petrol costs £1.80, determine the appropriate total cost (hiring cost plus petrol cost) of
(a) hiring a Range Rover for 7 days and travelling 450 miles,
(b) hiring a Nissan Cherry for 1 day and travelling 220 miles,
(c) hiring an Escort XR3i for 2 days and travelling 540 miles. (What a lot of driving!)

Solution
(a) There is no excess mileage charge when hiring for 7 days, so the hiring charge for this Group H vehicle is £600 plus VAT, where VAT will be 15% of £600

$$= \frac{15}{100} (600) = 90$$

Total hiring charge = 600 + 90 = £690.
The number of gallons of petrol required is $\frac{450}{15} = 30$,

giving petrol costs of 30(1.80) = £54.

Overall costs = 690 + 54 = £744.

(b) There is no excess mileage charge when hiring this Group B vehicle, so the hiring charge, in £, is

$$21.50 + \text{VAT} = 21.50 + \frac{15}{100}(21.50) = 21.50 + 3.22$$
$$= £24.72.$$

The petrol costs are $\frac{220}{30}(1.80) = £13.20$

Overall costs = £37.92.

(c) The hiring charge, ignoring VAT, is

2(39.50) + excess mileage charge
= 79 + 440(0.17) = £153.80,

as this Group E vehicle is subject to 17p extra for each mile greater than 100 miles.
Including VAT the hiring charge is

$$153.80 + \frac{15}{100}(153.80) = 153.80 + 23.07$$
$$= £176.87$$

The petrol costs are $\frac{540}{28}(1.80) = £34.71$.

Hence the overall costs = 176.87 + 34.71 = £211.58.

Example

Table 6.6 shows the number of divorcing couples in the years 1970 to 1976 broken down by the number and age of their children.
(a) What percentage of divorcing couples had no children in 1970 and in 1976?
(b) By what percentage had the number of divorcing couples increased from 1970 to 1976?

Solution
(a) In 1970 the percentage of divorcing couples with no children

$$= \left(\frac{22}{58}\right)100 = 37.9\%$$

Table 6.6 Divorcing couples by number and age of children – England and Wales (Thousands)

	1970	1971	1972	1973	1974	1975	1976
Divorcing couples:							
With: No children under 16	22	32	52	42	45	47	49
1 child under 16	15	17	27	24	26	28	30
2 children under 16	13	15	24	24	26	28	31
3 or more children under 16	8	10	16	16	16	18	18
Total divorcing couples	58	74	119	106	114	120	127
Children under 16 of divorcing couples:							
Ages of children:							
Under 5 years	19	21	30	30	32	33	34
5 years but under 11	35	41	64	62	65	69	71
11 years but under 16	18	21	36	36	39	43	47
Total children under 16 of divorcing couples	71	82	131	127	135	145	152
Average number of children under 16 per couple divorcing	1.22	1.11	1.10	1.20	1.19	1.20	1.20

Source: Social Trends, 1979

In 1976 the percentage of divorcing couples with no children

$$= \left(\frac{49}{127}\right) 100 = 38.6\%$$

Hence there is little change over the six-year period.
(b) The actual increase in divorces = 127 − 58 = 69 (thousands).
The percentage increase in divorces

$$= \left(\frac{69}{58}\right) 100 = 119\%,$$

as this is more than 100% it indicates that the number of divorces has more than doubled during this six-year period.

Many calculations can be done more easily and quickly by using mathematical tables. However, it should be remembered that the answers when using them may only be approximate.

Example
Table 6.7 can be found in many mathematical texts.
Use these tables to estimate, correct to three significant figures,
(a) $\sqrt{70}$
(b) $\sqrt{72.5}$

(c) n such that $\dfrac{1}{n} = 0.014$

Solution
(a) $\sqrt{70}$ can be obtained directly from column 4 of the table giving
$\sqrt{70} = 8.37$ to three significant figures
(b) $\sqrt{72} = 8.48528$
$\sqrt{73} = 8.54400$
As 72.5 is half-way between 72 and 73 then $\sqrt{72.5}$ will be approximately half-way between 8.48528 and 8.54400.
$$\sqrt{72.5} \simeq \frac{8.48528+8.54400}{2} = 8.51464$$

Hence $\sqrt{72.5} \simeq 8.51$

(c) We look at the final column of the tables to find the value of n
that gives the largest $\dfrac{1}{n}$ value smaller than 0.014.

This is $n = 72$, giving $\dfrac{1}{n} = 0.01389$.

Next we find the value of n that gives the smallest $\dfrac{1}{n}$ greate

Table 6.7 Squares, cubes, square roots, and reciprocals

n	n^2	n^3	$\sqrt{n}$	$\sqrt[3]{n}$	$\sqrt{10n}$	$\sqrt[3]{10n}$	$\sqrt[3]{100n}$	$1/n$
61	3 721	226 981	7.81025	3.93644	24.69818	8.48074	18.27107	0.01639
62	3 844	238 328	7.87401	3.95784	24.89980	8.52684	18.37037	0.01613
63	3 969	250 047	7.93725	3.97900	25.09980	8.57243	18.46861	0.01587
64	4 096	262 144	8.00000	4.00000	25.29822	8.61755	18.56581	0.01563
65	4 225	274 625	8.06226	4.02067	25.49510	8.66220	18.66201	0.01538
66	4 356	287 496	8.12404	4.04118	25.69047	8.70640	18.75722	0.01515
67	4 489	300 763	8.18535	4.06149	25.88436	8.75015	18.85148	0.01493
68	4 624	314 432	8.24621	4.08160	26.07681	8.79347	18.94481	0.01471
69	4 761	328 509	8.30662	4.10151	26.26785	8.83636	19.03722	0.01449
70	4 900	343 000	8.36660	4.12123	26.45751	8.87885	19.12875	0.01429
71	5 041	357 911	8.42615	4.14076	26.64583	8.92093	19.21941	0.01408
72	5 184	373 248	8.48528	4.16011	26.83282	8.96216	19.30922	0.01389
73	5 329	389 017	8.54400	4.17928	27.01851	9.00392	19.39820	0.01370
74	5 476	405 224	8.60233	4.19828	27.20294	9.04484	19.48637	0.01351
75	5 625	421 875	8.66025	4.21710	27.38613	9.08540	19.57376	0.01333
76	5 776	438 976	8.71780	4.23576	27.56810	9.12560	19.66037	0.01316
77	5 929	456 533	8.77496	4.25426	27.74887	9.16545	19.74622	0.01299
78	6 084	474 552	8.83176	4.27260	27.92848	9.20496	19.83133	0.01282
79	6 241	493 039	8.88819	4.29078	28.10694	9.24413	19.91572	0.01266

Source: Mathematical, Statistical, and Financial Tables for Social Sciences by Z. W. Kmietowicz and Y. Yannoulis

Table 6.8 Holiday prices per person in £'s

Hotel	Bahamas (full board)								Bahia de Palma (full board)							
Holiday Code	A13206								A13307							
No. of nights	7 Nights		10 Nights		11 Nights		14 Nights		7 Nights		10 Nights		11 Nights		14 Nights	
	Adult	Child	Adult	Child	Adult	Child	Adult	Child	Adult	Child	Adult	Child	Adult	Child	Adult	Child
21 Mar–31 Mar	194	111	202	114	208	117	222	122	193	136	199	136	204	138	219	143
1 Apr–7 Apr	172	99	194	102	199	105	214	109	171	117	192	127	197	129	211	134
8 Apr–30 Apr	149	88	174	99	181	102	197	105	139	102	164	116	171	118	186	123
1 May–18 May	159	112	181	124	187	126	206	134	147	109	166	119	173	121	188	128
19 May–26 May	187	134	198	137	204	139	226	149	175	131	182	133	187	135	207	139
27 May–19 Jun	177	125	202	135	209	137	231	147	165	116	187	123	192	125	212	132
20 Jun–3 Jul	181	127	205	139	212	142	235	152	182	129	207	141	231	155	259	167
4 Jul–17 Jul	194	134	221	147	239	154	267	166	207	147	242	165	254	171	287	187
18 Jul–24 Jul	229	156	267	171	277	176	309	189	229	162	267	181	279	187	312	203

25 Jul–17 Aug	238	165	271	179	282	184	312	197	237	171	272	189	284	194	317	211
18 Aug–24 Aug	235	157	268	171	279	176	306	187	235	163	267	181	281	187	307	201
25 Aug–31 Aug	227	152	252	163	265	168	288	178	225	158	249	171	264	177	286	188
1 Sep–7 Sep	212	145	238	156	247	159	272	171	207	151	235	162	244	167	269	181
8 Sep–21 Sep	209	143	236	154	245	158	269	168	205	147	232	161	241	164	265	176
22 Sep–28 Sep	199	138	222	148	234	153	251	159	194	141	214	149	227	156	236	161
29 Sep–16 Oct	166	116	192	127	199	129	219	138	156	113	179	123	186	126	202	133
17 Oct–24 Oct	164	119	189	129	197	133	217	142	159	116	182	126	189	129	205	136

Source: IntaSun Holiday Brochure, 1986

than 0.014 which is $n = 71$, and $\dfrac{1}{n} = 0.01408$.

The number that needs to be added to 0.01389 to make 0.014 is 0.00011. The total difference between 0.01389 and 0.01408 is 0.00019.

An estimate of n that gives $\dfrac{1}{n} = 0.014$ is

$$72 - \frac{0.00011}{0.00019} = 72 - \frac{11}{19} = 71\tfrac{8}{19} = 71.4$$

What answer does your calculator give?

Example

Table 6.8 comes from a holiday brochure distributed by a leading travel company in Spring 1986.

A family of two adults and three children are considering two possible holidays:

(a) 10 nights at the Hotel Bahamas departing on 6 May,
(b) 7 nights at the Hotel Bahia de Palma departing on 20 July.

Which of these two holidays is the cheaper?

Solution

(a) The cost of one adult departing on 6 May is £181 for 10 nights and the corresponding cost per child is £124.
    The total cost $= 2(181) + 3(124) = £734$
(b) Departing on 20 July, a 7 night holiday to the Hotel Bahia de Palma costs £229 per adult and £162 per child, giving a total cost of

$$2(229) + 3(162) = £944$$

The 10-night holiday in May is more than £200 cheaper than the 7-night holiday in July.

Exercises

1. Table 6.9 shows the monthly repayments to pay off a mortgage loan over certain time periods when the interest rate is 12.75% gross (8.925% net).
   Assuming an interest rate of 8.925% net, use this table to estimate the monthly repayments for the following mortgages:

Table 6.9 Calendar monthly repayments at MIRAS rate 8.925% net (Gross rate 12.75%)

Amount advanced	Term of years				
	10	15	20	25	30
	£	£	£	£	£
£10	0.13	0.10	0.09	0.08	0.08
£50	0.65	0.52	0.46	0.42	0.40
£100	1.30	1.03	0.91	0.85	0.81
£200	2.59	2.06	1.82	1.69	1.62
£300	3.89	3.09	2.73	2.54	2.42
£400	5.18	4.12	3.64	3.38	3.23
£500	6.48	5.16	4.55	4.23	4.04
£600	7.78	6.19	5.46	5.07	4.85
£700	9.07	7.22	6.37	5.92	5.66
£800	10.37	8.25	7.28	6.76	6.46
£900	11.66	9.28	8.19	7.61	7.27
£1 000	12.96	10.31	9.10	8.45	8.08
£2 000	25.92	20.62	18.20	16.90	16.16
£3 000	38.88	30.93	27.30	25.35	24.24
£4 000	51.84	41.24	36.40	33.80	32.32
£5 000	64.80	51.55	45.50	42.25	40.40
£6 000	77.76	61.86	54.60	50.70	48.48
£7 000	90.72	72.17	63.70	59.15	56.56
£8 000	103.68	82.48	72.80	67.60	64.64
£9 000	116.64	92.79	81.90	76.05	72.72
£10 000	129.60	103.10	91.00	84.50	80.80

Table 6.10 Insurance premiums (£)

Vehicle group	Age	No claims bonus					
		Max	4–6 years	3 years	2 years	1 year	Start
1	45–70	23	25	31	37	41	46
	30–44	25	27	34	40	45	50
	28–29	28	30	37	45	49	56
	25–27	30	31	39	47	52	58
	23–24	32	34	43	51	57	68
2	45–70	24	25	32	38	42	48
	30–44	27	28	35	42	47	53
	28–39	30	31	39	47	52	58
	25–27	30	32	40	48	53	60
	23–24	34	35	44	53	59	71
3	45–70	30	31	40	47	53	59
	30–44	33	35	43	52	58	65
	28–29	37	39	48	58	64	72
	25–27	38	40	50	60	67	75
	23–24	42	44	55	66	73	88
4	45–70	34	36	44	53	59	67
	30–44	37	39	49	58	65	73
	28–29	41	43	54	64	71	80
	25–27	42	45	56	67	74	83
	23–24	47	49	61	74	82	92
5	45–70	41	43	54	65	72	81
	30–44	45	48	59	71	79	89
	28–29	50	53	66	79	88	99
	25–27	52	55	68	82	91	102
	23–24	57	60	75	90	100	120
6	45–70	45	47	59	71	79	88
	30–44	49	52	64	77	86	97
	28–29	54	57	71	85	95	107
	25–27	56	59	74	89	98	111
	23–24	62	65	81	97	108	130

Source: What Car? Apr. 1986

(a) A mortgage of £8000 repaid over 20 years.
(b) A mortgage of £24000 repaid over 30 years.
(c) A mortgage of £7500 repaid over 10 years.
(d) A mortgage of £6000 repaid over 12 years.

2. Table 6.10 shows the car insurance rates for 12 months third party, fire and theft insurance as published by an insurance broker.
Determine the cost of insuring
   (a) a group 4 vehicle if the insured is aged 43 and has 4 years no claims bonus,
   (b) a group 3 vehicle if the insured is aged 24 and has 2 years no claims bonus.

3. Table 6.11 shows a list of marks (out of 100) gained by 60 students in an examination. Arrange them in a grouped frequency table.

Table 6.11 Examination marks

78	15	59	66	22	41	44	12	84	32
56	81	50	45	82	58	75	34	69	61
67	55	52	73	31	48	41	56	91	56
84	66	56	44	39	45	51	68	70	64
66	27	47	78	24	53	32	24	72	53
59	42	78	64	76	38	87	41	54	36

4. Table 6.12 shows the number (in thousands) of births in the United Kingdom.
   (a) Determine the total number of births in the United Kingdom over the years 1980 to 1984 inclusive.
   (b) For each of the four years 1981 to 1984 determine the percentage of births that occur in the third quarter of the year.

Table 6.12 Births in the United Kingdom 1980–85

	United Kingdom	England and Wales		Scotland	Northern Ireland
Live births*					(thousands)
		Total	Wales		
1980	753.7	656.2	37.4	68.9	28.6
1981	730.8	634.5	35.8	69.1	27.3
1982	719.2	625.9	35.7	66.2	27.0
1983	721.5	629.1	35.5	65.1	27.3
1984	729.6	636.8	35.9	65.1	27.7
1981 1st quarter	180.5	156.2	8.9	17.5	6.7
2nd quarter	184.9	160.8	9.2	17.1	7.0
3rd quarter	188.5	164.3	9.2	17.2	7.0
4th quarter	176.9	153.1	8.5	17.2	6.6
1982 1st quarter	176.7	153.4	8.6	16.6	6.7
2nd quarter	180.2	157.0	9.0	16.3	6.9
3rd quarter	185.9	162.1	9.3	16.7	7.0
4th quarter	176.5	153.4	8.8	16.7	6.4
1983 1st quarter	175.3	152.4	8.7	16.1	6.8[†]
2nd quarter	184.6	161.3	9.0	16.3	7.0[†]
3rd quarter	187.4	163.4	9.1	16.9	7.1[†]
4th quarter	174.1	151.9	8.6	15.8	6.3[†]
1984 1st quarter	176.1	153.5	8.6	15.7	6.9[†]
2nd quarter	180.7	157.7	8.8	15.9	7.0[†]
3rd quarter	191.0	167.1	9.4	16.9	7.0[†]
4th quarter	181.8	158.4	9.0	16.7	6.7[†]
1985 1st quarter	183.2	160.3	9.0	16.0	6.9[†]
2nd quarter	190.0[‡]	166.0[‡]	9.0	16.6	7.1[†]
3rd quarter		173.0[‡]			7.2[†]

* Figures for England and Wales relate to occurrences. Figures for Scotland and Northern Ireland relate to births registered.
[†] Provisional
[‡] Estimated from births registered in the period.
Source: Monthly Digest of Statistics, Nov. 1985

Unit 7
Charts

The main purpose of a chart or diagram is to convey information simply, clearly, quickly, and in such a way that merely presenting the raw data in a table does not do. There are a few characteristics which all charts or displays of data must have.
1. There must be a title.
2. All units of measurement must be clearly stated.
3. There must be a key to any symbols or shading used.
4. The source of the data must be given if it is not obvious.
5. Any ambiguous or non-standard terms used must be defined, usually in a footnote.
6. The diagram must be sufficiently large for the detail to be clear.

Consider the chart shown in Fig. 7.1, which clearly shows that the percentage of female employees in total employment rose

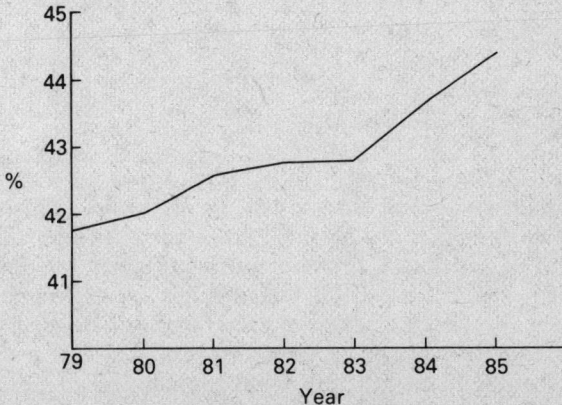

Fig. 7.1 Proportion of female employees in total employment (*Source: Economic Outlook*, vol. 10, No. 1)

steadily from 1979 to 1983 and then showed a more marked increase between 1983 and 1985. This chart satisfies all the requirements of a good chart, and conveys the underlying pattern contained in the raw data much better and quicker than the data alone could have done.

Figure 7.2 is another example of a clear diagram which shows the enormous increase in adult unemployment since 1980. The slight drop in unemployment after 1985 is a forecast, not the truth, and this is made clear by the diagram. Both of the previous charts showed how a particular quantity varied over time, and as such are examples of *time series*, and the convention when displaying any time series is to have time on the horizontal axis as in Fig. 7.1 and 7.2.

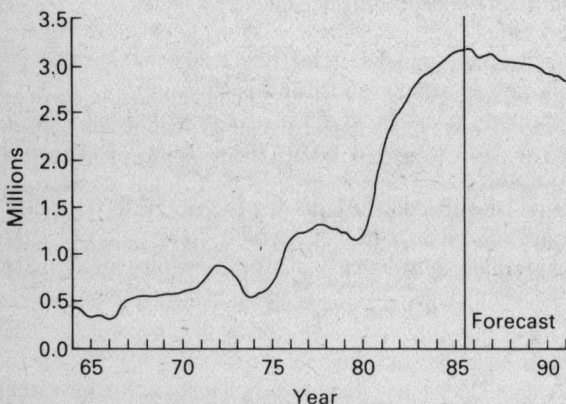

Fig. 7.2 Adult unemployment
(*Source: Economic Outlook,* vol. 10, No. 1)

One of the commonest types of chart is called a *histogram*. Figure 7.3 is a histogram showing how the heights of men vary. As with all histograms the area of each block represents the frequency for the interval on which it stands, and if all the intervals are the same width, as they are here, then the height of each block can be taken to represent the frequency. It can be seen from Fig. 7.3 that few men are smaller than 160 cm (5ft 3in) or taller than 185 cm (6ft 1in) and the most 'popular' or common height is just less than 175 cm (5ft 9in). Notice that the footnote gives further information on the men measured.

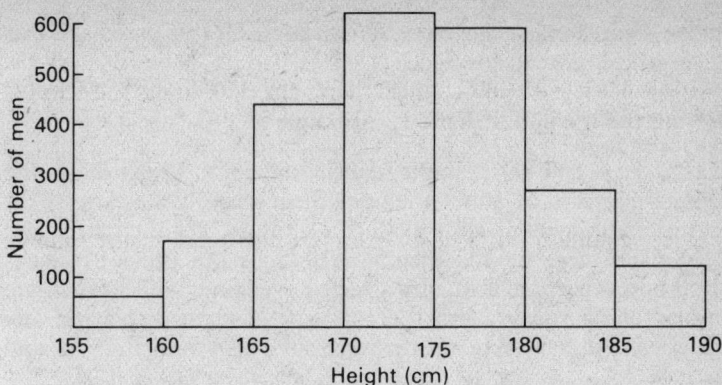

Fig. 7.3 Height distribution of men[1] aged 16–64
1. Sample of 4 514 men taken in 1980
(*Source: Adult Heights and Weights Survey*, OPCS)

Figure 7.4 is a good example of a *pie-chart*, where a circle or disc is divided into segments whose size, as measured by the area (and hence the angle at the centre) represents the proportion or percentage associated with the label of the segment. It can be seen from this pie-chart that nearly twice as much is spent on food than on clothing and footwear, while 'Other services' account for the largest proportion of expenditure. (What do you think would come under the heading "Other services"?) In a pie-chart it is often difficult to judge which of two segments is the smallest, and so the actual figures should always be given to allow for such comparisons. Here the diagram does not make it clear that the expenditure on durable goods is slightly higher than the expenditure on other goods, but

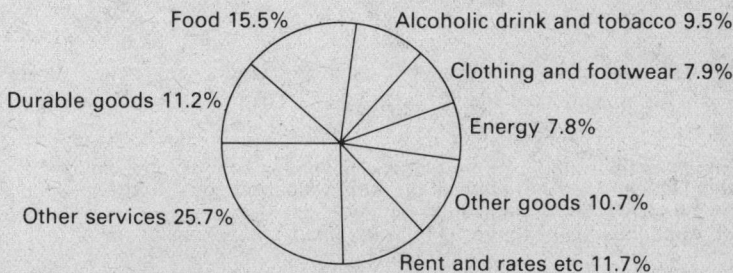

Fig. 7.4 Share of consumers' expenditure 1984 (1980 prices)
(*Source: Economic Trends*, No. 385)

the given figures make this clear. When drawing this pie-chart, the angle of the segment for food (15.5%) had to be evaluated by calculating 15.5% of 360°, since there are 360° in a full circle. Following the method of Unit 3, this angle is

$$\frac{15.5}{100} \times 360 = 55.8°.$$

Other, less common types of diagram are illustrated in the following examples.

Example

The chart shown in Fig. 7.5 is an example of a *multiple bar chart*, similar to a histogram in that height represents frequency or

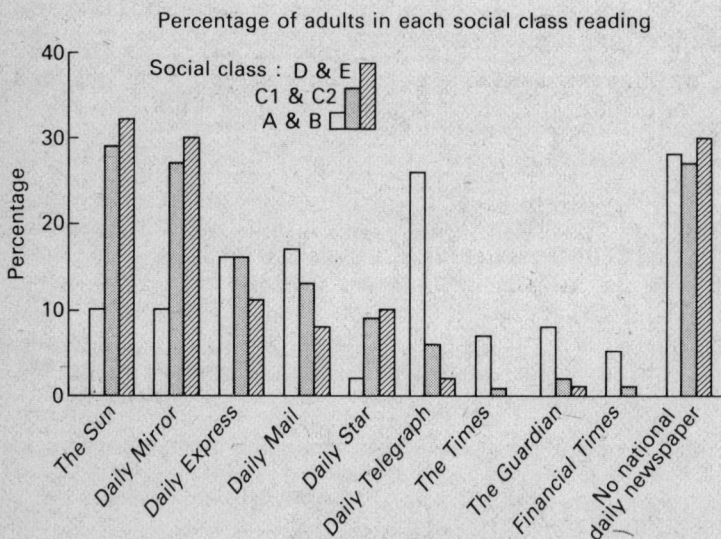

Percentage of adults in each social class reading

[1] Defined as the number of people who have read or looked at one or more copies of a given publication during a period equal to the interval at which the publication appears.

[2] See Appendix, Major Surveys: *National Readership Survey*.

Fig. 7.5 Reading of national newspapers[1] by social class[2], 1980 (Great Britain)
(*Source: Social Trends*, No. 12, 1979)

popularity but where the variable along the horizontal axis is not one measured on a continuous scale such as height or age. The social class divisions used here are the standard ones used based upon occupation, and are defined precisely in the Appendix mentioned in footnote 2, with A and B being upper class, C1 and C2 middle class, and D and E working class.

(a) Of the newspapers mentioned, which is read by the fewest people?
(b) Which newspaper is read by the most people?
(c) Which newspapers could be called 'upper class'?
(d) Which newspapers could be called 'working class'?

Solution

To find the percentage of people who read any particular newspaper, the three percentages for each of the class divisions need to be averaged, and this is not really straightforward since there are more people in social classes D and E than in A and B. However, despite this complication, the following answers are clear from the diagram:

(a) *The Times*, *Guardian* and *Financial Times* have the fewest readers, with the *Financial Times* having the least.

(b) The *Sun* has the most readers, with the *Daily Mirror* a close second.

(c) Besides *The Times*, *Guardian* and *Financial Times*, the *Daily Telegraph* shows a clear 'upper class' readership. For classes A and B the *Daily Telegraph* is the most read newspaper, but it can be seen that in classes A and B more people do not read any newspaper than read the *Daily Telegraph*!

(d) The *Sun* is read by more people in classes D and E than any other newspaper, but is still read by approximately 10% of people in classes A and B, so perhaps the *Daily Star* should be called 'working class' because it shows the greater imbalance. No precise answer can be given without a better explanation of how to judge whether a newspaper is 'working class'.

Example

Figure 7.6 shows the age structure of the United Kingdom in 1980. It is not a common type of chart, but does allow for many interesting observations. It clearly shows that for either sex the number of people alive in 1980 who were born before 1900, i.e. aged over 80 in 1980, is very low, as expected. Study this chart

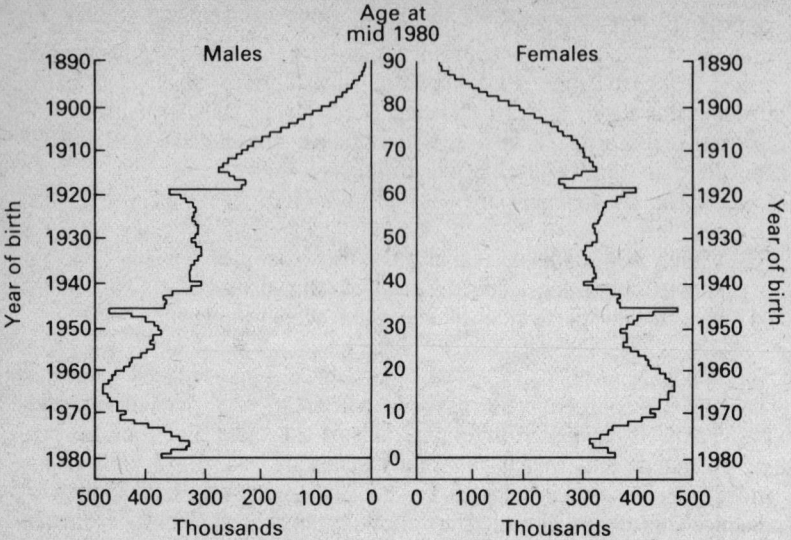

Fig. 7.6 Population: by age and sex, 1980 (United Kingdom)
(*Source: Social Trends*, No. 12, 1979)

carefully and then make at least three other comments that this chart indicates.

Solution
(a) Females tend to live longer than males, since there are many more old ladies than old men.
(b) There was a sharp increase in the birth rate, for both sexes, in 1920, probably caused by men returning home after the First World War.
(c) There was a similar increase in the birth rate after the Second World War.
(d) There was a decrease in the birth rate during 1916–19, caused by the absence of men away from home fighting in the First World War.
(e) The fall in the birth rate during the Second World War was not as great as it had been in the First World War. This may have been due to better transportation which allowed more frequent leave, or because there were more men in the armed forces stationed in Britain!

(f) There was a 'bulge' in the birth rate around 1965, which has produced a 'bulge' in the number of people of both sexes who were 15 years old in 1980.

Exercises

1. Study Fig. 7.7, and then answer the following questions:

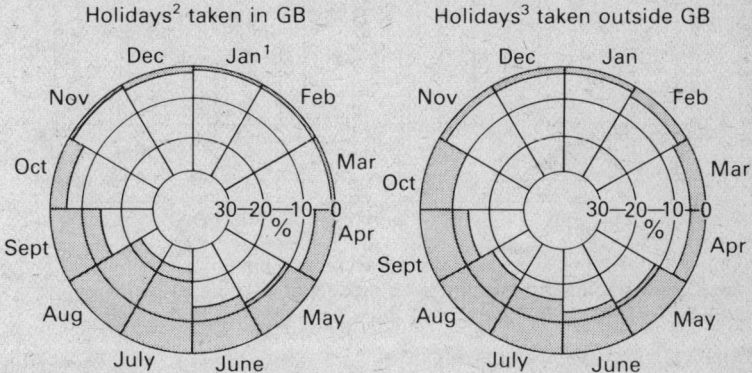

Holidays[2] taken in GB   Holidays[3] taken outside GB

[1] Month in which holiday started.   [3] Holidays of 1 or more nights.
[2] Holidays of 4 or more nights.

Fig. 7.7 Monthly distribution of holidays taken by adults resident in Great Britain, 1980
(*Source: Social Trends*, No. 12, 1979)

   (a) For holidays taken in Great Britain, which months are the most popular, and which the least popular?
   (b) For holidays taken outside Great Britain which month is the most popular and which months the least popular?
   (c) Is it true to say more people take August holidays in Great Britain than outside?

2. Study Fig. 7.8, and then answer the following questions, which refer only to people who do drink alcohol (see footnote 2).
   (a) For men, which age group shows the highest consumption of alcohol and which the lowest?
   (b) For women, which age group shows the highest consumption of alcohol and which the lowest?
   (c) Does this chart confirm the statement that on average men drink twice as much as women?

Average alcohol consumption during week preceding interview

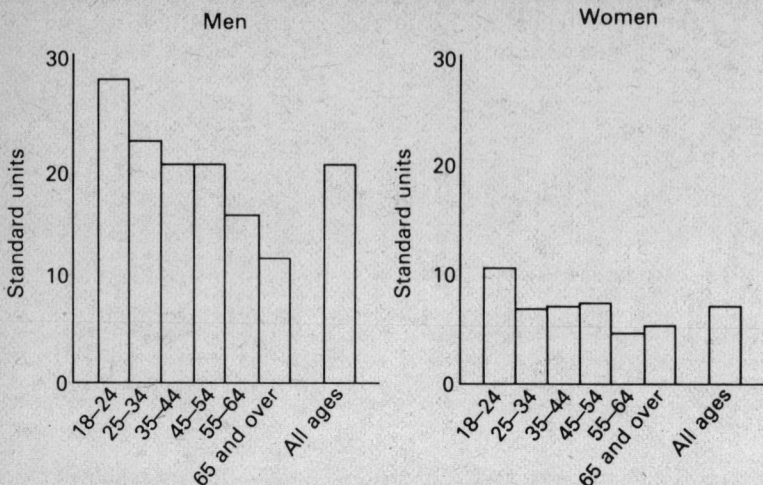

Fig. 7.8 Alchohol consumption[1&2] of drinkers by sex and age, 1978 (England and Wales)
(*Source: Social Trends*, No. 12, 1979)

[1] One standard unit is equivalent to half pint of beer.
[2] Excludes people who had nothing to drink during the year preceding interview.

3. Study Fig. 7.9, and then answer the following questions:
   (a) For females, which age group spend the least time on drink-associated activities and the most time watching TV?
   (b) Would your answer to (a) be the same for males?
   (c) Describe briefly the main differences in leisure activities between males and females in the 18–24 age group.
   (d) Is there a link between the facts shown in this chart and those shown in Fig. 7.8?

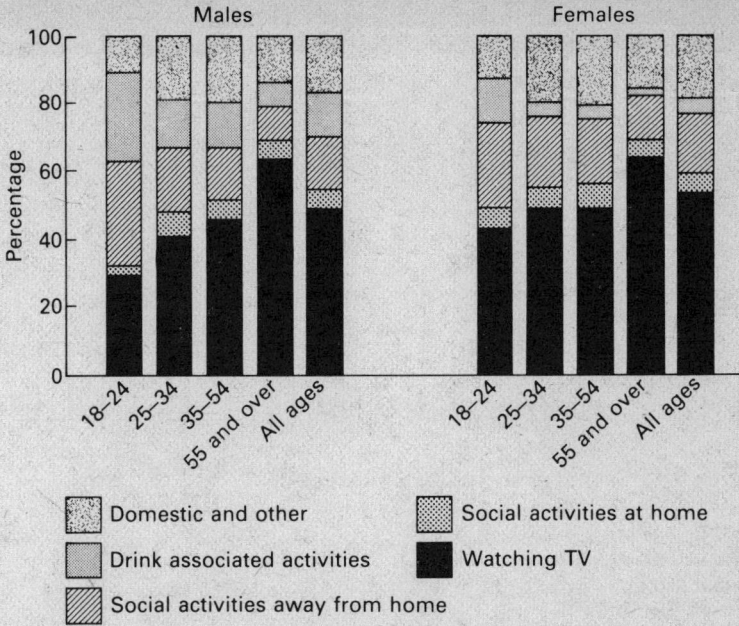

Males   Females

¹ The proportions show the average percentage of evenings, during the 7 days before interview, on which people said they spent most of the evening doing the specified activity.

Fig. 7.9 Proportion of evenings¹ spent on various leisure activities by sex and age, 1978 (England and Wales)
(*Source: Social Trends*, No. 12, 1979)

Unit 8
Co-ordinates

Table 8.1 shows the scores in six first-division football matches played one Saturday in December 1984.

Table 8.1   Scores in six football matches

Match number	(i)	(ii)	(iii)	(iv)	(v)	(vi)
Home team	1	1	1	3	2	0
Away team	0	2	1	1	0	1

One convenient method of illustrating this data is to use a co-ordinate system in which each match is represented by a point or a dot. The position of this point is determined by the number of goals scored by the two sides. A suitable co-ordinate system is shown in Fig. 8.1 where the six matches or points have been placed in the appropriate position. For example, match number (iv) is represented by a point which has the co-ordinates Home score 3, Away score 1. The point when both teams score 0 is called the *origin*.

Do these six results show a trend? Not clearly, but perhaps more matches would show that the home team tends to score more goals than the away team. You might try this.

The data used in this example were nice and simple (only the values 0,1,2,3 were used), but a co-ordinate system together with suitable axes can be used to display data of any magnitude. The data shown in Table 8.2 were taken from seven adults, and are an attempt to investigate the relationship between height and head circumference measured around the head just above the eyes.

The two axes which should be used to display this data are height and head circumference, but since all heights are greater

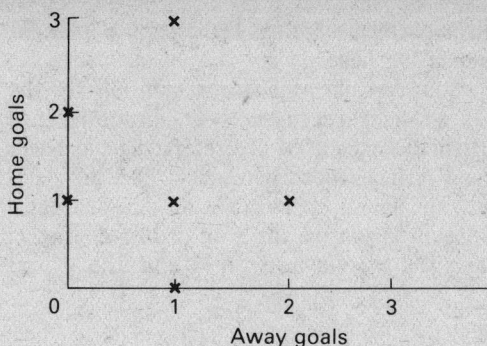

Fig. 8.1 Results of six football matches

Table 8.2 Height and head circumference of seven adults

Person	A	B	C	D	E	F	G
Height (cm)	183	185	175	168	166	165	161
Head circumference (cm)	61	59	58	57	56	62	55

than 160 cm, there is no need to start the height axes at 0 cm since this would waste a lot of space. One possible method is to start the height axes at 150 cm and the head circumference axes at 50 cm, as is shown in Fig. 8.2.

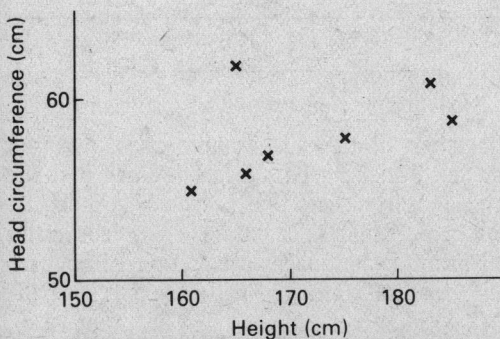

Fig. 8.2 Height and head circumference for seven adults

75

These points show a clear pattern which is broken by person F, who either has short legs or a big head!

It is common in many situations like this to use symbols for the two variables involved. If we let y represent head circumference and x represent height, then the axes used are referred to as the y axis and x axis, and person A has co-ordinates $x = 183$, $y = 61$. This is often abbreviated by saying person A has co-ordinates (183,61), where the convention is to give the x co-ordinate first. It is also convention to have the x axis horizontal and the y axis vertical.

Example

It is clearly reasonable to expect some relationship between head circumference and height. One suggested relationship is that height is three times head circumference. Test this relationship using the head circumference of the seven people discussed earlier.

Solution
The first step is to calculate the expected height if the relationship is true, i.e. multiply each head circumference by three. This produces the results in Table 8.3.

Table 8.3 Head circumference and predicted height of seven adults

Person	A	B	C	D	E	F	G
Head circumference (cm)	61	59	58	57	56	62	55
Height predicted (cm)	183	177	174	171	168	186	165

These seven points are then plotted on suitable axes, together with the original points. See Fig. 8.3. This clearly shows that the claim is reasonable, but again emphasises that person F is unusual.

The final example of co-ordinates is taken from computer graphics. Most computer graphics languages use a conventional co-ordinate system with a horizontal x axis, a vertical y axis and the notation that the point (2,3) is at $x = 2$, $y = 3$. The point (0,0), usually called the origin, is situated at the bottom left-hand corner

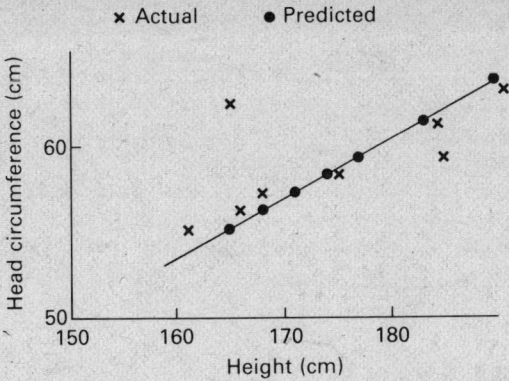

Fig. 8.3 Actual compared to practical head circumferences

of the screen. The two commands used in one particular system are:

MOVE X,Y
which moves the imaginary pen or cursor, to the point (X,Y) but no line or point is drawn, and
DRAW X,Y
which causes a line to be drawn from the current position of the cursor to the new position X,Y.

Example

What shape would the following commands produce?

MOVE 10,10
DRAW 110,10
DRAW 110,60
DRAW 10,60
DRAW 10,10
MOVE 10,60
DRAW 60,80
DRAW 110,60

Solution

First a co-ordinate system is needed with *x* and *y* axes, with *x* varying from 0 to 110 and *y* varying from 0 to 80. The results of following the above commands using graph paper are shown in Fig.

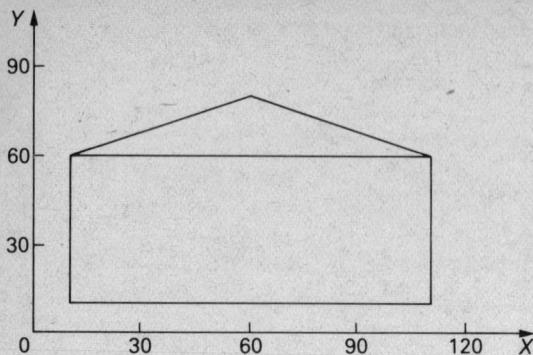

Fig. 8.4 Computer graphics example

8.4, but on the monitor screen only the house shape would be drawn; no axes would be seen.

Example

This example is not applied to any particular practical problem, but is one which will often occur in more advanced areas of study.

Using a suitable co-ordinate system illustrate graphically the relationship between y and x if $y = 2x + 1$.

Use values of x between -4 and $+4$.

Solution

We shall evaluate pairs of numbers (x,y) so that each pair satisfies this relationship. For example if $x = 3$, then $y = 2(3) + 1 = 6 + 1 = 7$. Each pair, such as $(3,7)$ will then represent a point in a co-ordinate system. The results of evaluating some of these pairs are shown in Table 8.4.

Table 8.4

x	-4	-3	-2	-1	0	1	2	3	4
y	-7	-5	-3	-1	1	3	5	7	9

(If you are not sure about negative numbers read Unit 1)

The result of plotting these points can be seen in Fig. 8.5, which shows the nature of the relationship. In Fig. 8.5 the lowest point has co-ordinate $(-4,-7)$ and the highest has co-ordinate $(4,9)$.

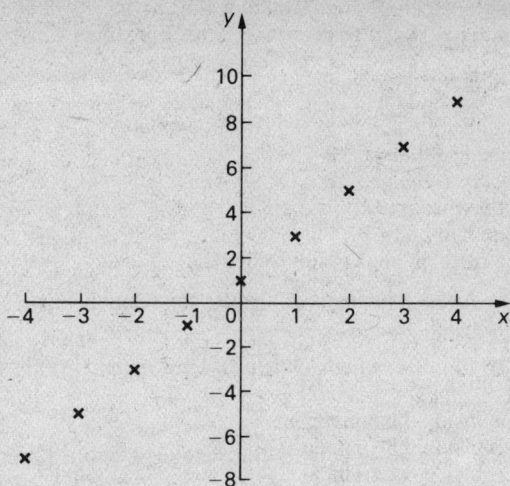

Fig. 8.5 The relationship $y = 2x + 1$

Exercises

1. Ten car drivers were each asked how many cars they had owned and how long they had held a full licence, with the following results:

Person	A	B	C	D	E	F	G	H	I	J
No. of cars	6	2	8	4	1	1	15	7	9	3
No. of years	10	1	15	23	1	2	4	10	17	4

Choose suitable axes, display this data, and briefly comment.

2. Using axes as in Fig. 8.1 plot the results of five imaginary football matches in which the home team always won.

3. It is claimed that there is a strong relationship among women between the size of their shoes and their hip measurements. Test this claim by measuring, recording and then plotting the data from a sample of six women. Would you expect the same relationship among men?

4. What shape would be produced if the following instructions were obeyed by a computer with a graphics monitor?
MOVE 20,100
DRAW 80,100
MOVE 50, 30
DRAW 50,140
MOVE 45,140
DRAW 55,140

5. Using x and y co-ordinates which both go from 0 to 150, write instructions for a computer so that a large capital letter F is drawn.

6. The six points A,B,C,D,E and F have (x,y) co-ordinates as follows:
   A: (1,3)   B: (4,2)   C: (2,1)
   D: (7,2)   E: (3,6)   F: (4,4)
   (a) Which point is nearest to the x axis?
   (b) Which point is nearest to the y axis?
   (c) Which point is furthest from (0,0), the origin?
   (d) Which point is nearest to point B?
   (You may need to use a ruler on your graph to decide.)

7. Using values of x from -5 to $+5$ inclusive, illustrate the relationship between y and x if $y = 4x - 2$.

8. Decide with reference to Fig. 8.4, whether the following statements are true or false:
   (a) The point (40,30) is inside the rectangle.
   (b) The point (80,65) is inside the triangle.
   (c) The point (20,70) is inside the triangle.
   (d) The straight line from (80,20) to (0,80) cuts into the triangle.
   (e) The straight line from (10,10) to (110,60) divides the rectangle into two triangles of equal area.

9. Write instructions for a computer so that the word HEALTH is drawn. You can assume that the x and y co-ordinates both vary from 0 to 150.

Unit 9
Graphs

Like the charts discussed in Unit 7, one of the functions of a graph is to display data so that any pattern, trend or relationship can easily be seen. A graph, however, is a specific type of diagram which uses a co-ordinate system as described in Unit 8 to show the relationship between two variables. The two variables may be related precisely, which is the case with an *equation* connecting the two variables (Unit 11) or may have what is called an *empirical* relationship. An empirical relationship is one that is not always exactly true, but tends to be approximately true and is based upon many observations of the two variables. One example of an empirical relationship is that for normal adults their height is approximately three times the circumference of their heads measured around the forehead (see Unit 8).

Is there an empirical relationship between a person's height and weight? The first step in investigating such a relationship is to collect heights and weights of a large sample of people. Table 9.1 shows some of the results of such a survey.

Table 9.1   Heights and weights of women aged 16 to 64, 1980

Height group (cm)	150.1–155.0	155.1–160.0	160.1–165.0	165.1–170.0	170.1–175.0
Group midpoint (cm)	152.55	157.55	162.55	167.55	172.55
Average weight (kg)	57.3	60.5	62.8	65.1	68.7

Source: Adult Height and Weight Survey, OPCS

The basic relationship is clear from this table; that is, the taller a woman is then, on average, the heavier she is, which is only what would have been expected. Any further comments and conclusions can best be made after a suitable graph is drawn showing the

81

relationship between these two variables. Let the variable H stand for the midpoint of the height group, and W stand for the average weight of those women in any particular height group. Since the weight is the variable we are interested in, that is, to be predicted from height, it is a convention to choose a co-ordinate system with weight W, on the vertical axis and height H, on the horizontal axis. More formally W is the *dependent* variable which depends upon the *independent* variable H.

Having decided which way round to have the axes, the next problem is to decide upon a scale which will allow the graph to be drawn on the piece of paper available (preferably graph paper with 1 cm squares). The horizontal variable, or independent variable, H varies from 152.55 cm to 172.55 cm, but to allow a little leeway at each end we can have H varying from 150 cm to 180 cm. To start H at zero would produce a tiny graph and a great deal of wasted paper! The weight axes can be started at 55 kg and go up to 70 kg, but since the data to be plotted is the average weight, it would be better to let the weight vary from 40 kg to 90 kg so that any individual, overweight or underweight, can be plotted on the graph. Using these axes and the data in Table 9.1, the graph in Fig. 9.1 can be drawn, where the points, when plotted, are joined

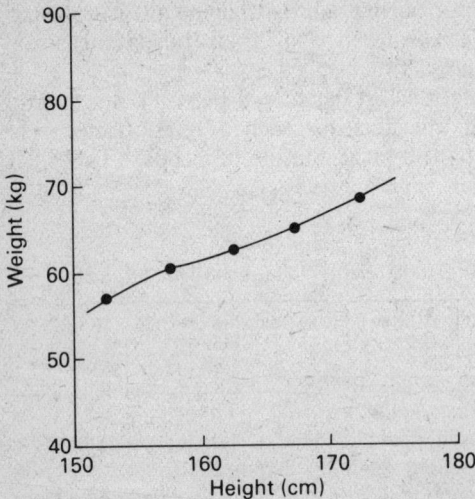

Fig. 9.1 Average weight (W) and height (H) of women aged 16–64 1980

(*Source: Adult Height and Weight Survey*, OPCS)

together to show the relationship between height and average weight. The curve joining the points should be reasonably smooth, and pass through all the points if possible.

One advantage of this graph over the tabulated data is that the average weight for any particular height can be found. For example, women who are 165 cm tall (5 ft 5 in) weigh, on average, 63.5 kg (10 stone) and so a woman of 165 cm who weighs 70 kg (11 stone) could be said to be 6.5 kg (1 stone) overweight when compared with other women of her height.

The *gradient* of a curve is the rate of change of the variable on the vertical axis with respect to the variable on the horizontal axis. If a graph is drawn with distance travelled on the vertical axis and time taken to travel this distance on the horizontal axis, then the gradient at any point is the rate of change of distance with respect to time; that is, speed. Consider Fig. 9.2 which shows the progress of a canal barge between two locks 300 m apart. The gradient at the start and end of the journey is zero (flat), showing that the barge started from rest and stopped when it reached the next lock. For the first minute it picked up speed and then maintained this steady speed for a further two minutes. The next two minutes were spent at a slower speed before slowing down and finally stopping after a total journey of 300 m.

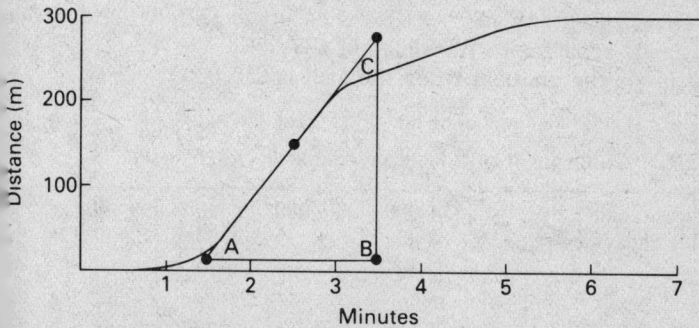

Fig. 9.2   Distance, time for a canal barge

The gradient or slope of a curve at any particular point can be estimated from a graph by first drawing a tangent to the curve at that point, i.e. a straight line that just touches the curve but does not cut it, and then measuring the gradient of this tangent. The gradient of a straight line is the ratio of the increase along the vertical axis to the corresponding increase along the horizontal axis.

To estimate the gradient or speed of the barge after 2 minutes, draw a tangent to the curve at 2 minutes, as in Fig. 9.2. This line has gradient

$$\frac{CB}{AB} = \frac{265}{2} = 132.5 \text{ m/min.}$$

This is equivalent to just less than 8 km/hour, or a very fast walk. In the examples that follow, gradients are further investigated with examples of zero gradients (flat) and negative gradients (curves going down instead of up).

Example

One of the commonest methods of buying a house is to have a repayment mortgage where the loan initially borrowed to buy the house is repaid over a fixed period of years by equal monthly repayments. Table 9.2 shows the monthly repayment needed to repay varying loans over a 25-year period when the net rate of interest is 8.925% p.a.

(a) Draw a graph to show the relationship between the size of loan and the monthly repayment. (The size of loan axis should go from zero to £40,000.)
(b) Use your graph to estimate the monthly repayments required for a loan of £16,500.
(c) Extend the graph to cover loans of up to £40,000 and hence estimate repayments for a loan of £37,950.
(d) Estimate the gradient when the loan is £10,000.

Table 9.2 Calendar monthly repayments at 8.925% p.a.

Amount of loan (£)	1 000	5 000	10 000	20 000
Monthly repayments (£)	8.45	42.25	84.50	169.00

Solution
(a) See Fig. 9.3 which shows a clear straight line or *linear* relationship.
(b) A loan of £16,500 will need a monthly repayment of £13 approximately. (A more accurate answer is £139.43 which ca be found using the methods of Unit 11, in which this exampl is discussed.)
(c) The straight line can easily be extended as shown in Fig. 9.

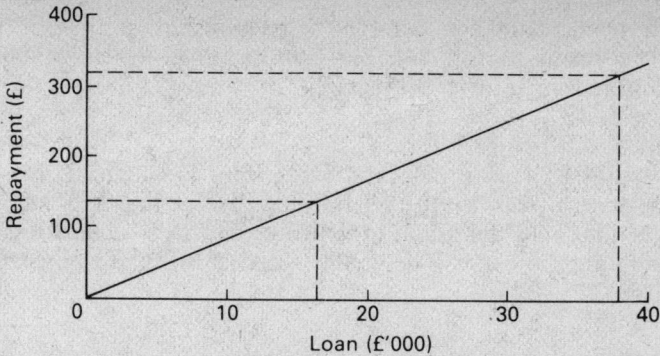

Fig. 9.3 Monthly repayments over 25 years at 8.925% p.a.

and this gives an estimate of £320.00. (Again the methods of Unit 11 produce a more accurate answer of £320.68.)

(d) The gradient = 84.50/10,000 = 0.00845. Any triangle will do, preferably a large one, since the graph is a straight line and has the same gradient all along it.

Example

A situation that faces many people is that of borrowing money to buy a car. How much can you afford to borrow? Table 9.3 is typical of many loan repayment tables and shows the various sums that can be borrowed over different repayment periods if the monthly repayment is fixed at £50.00.

Table 9.3 Loan and repayment term for a monthly repayment of £50.00

Repayment term (months)	Loan (£)
12	540.54
18	772.56
24	983.48
30	1176.47
36	1356.55
48	1621.53
60	1875.11

(a) Graph this data and comment on the relationship.
(b) Use the graph to estimate the loan possible for a £50.00 monthly repayment over $3\frac{1}{2}$ years.

Solution
(a) See Fig. 9.4. This is not a straight line, although it does demonstrate a clear, precise relationship. To see that it is not a straight line hold the graph to your eye and look along it, so exaggerating the curve. The gradient is decreasing as the term increases.

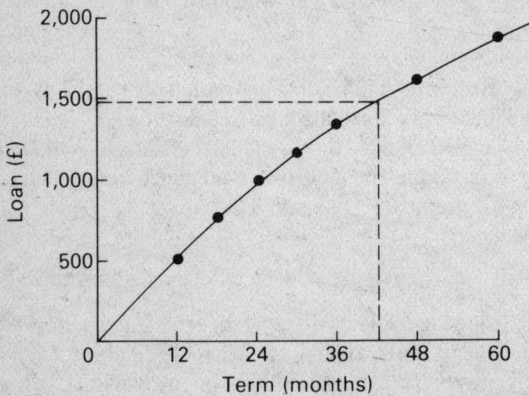

Fig. 9.4 Loan and repayment term for a monthly repayment of £50.00

(b) When the term is 42 months the initial loan is approximately £1480.00.

Example

Before mass-produced products such as transistors or light bulbs are allowed to leave the factory it is usual to have some form of inspection. If a batch fails the inspection, then all faulty or defective items are replaced by sound ones, and so the whole batch will contain no defectives. If a batch passes the inspection, it may still contain defectives since the inspection often involves testing a small sample. Table 9.4 shows the percentage of defectives leaving th

Table 9.4 Percentage defectives manufactured and outgoing after inspection

% Defectives manufactured	% Defectives outgoing
0	0.0
2	1.8
4	3.2
6	3.9
8	4.1
10	4.0
12	3.7
14	3.2
16	2.7
18	2.2
20	1.8
25	1.0
30	0.5
40	0.1

factory, after one type of inspection, and possible replacement, for various values of the percentage defective produced in manufacture.

(a) Graph this data and comment.

(b) What is the worst percentage defective that leaves the factory?

(c) Estimate the gradient when the percentage defective is (i) 3%, (ii) 6%, (iii) 15%.

Solution

a) Figure 9.5 shows a smooth curve which gets nearer and nearer to the horizontal axis. If no defectives are made, then obviously no defectives leave. If many of the items produced are defective then most inspections will fail, leading to all defectives being replaced, that is, there will again be few defectives leaving.

b) The curve clearly reaches a maximum of approximately 4.1%.

c) (i) When 3% defective, gradient is $\dfrac{CB}{AB} = \dfrac{3.0}{4.5} = +0.67$.

(ii) When 8% defective, gradient is 0.

(iii) When 15% defective, gradient is $-\dfrac{DE}{EF} = -\dfrac{3.2}{12} = -0.27$

This last gradient is negative because as the % defective increases, the outgoing defectives decrease.

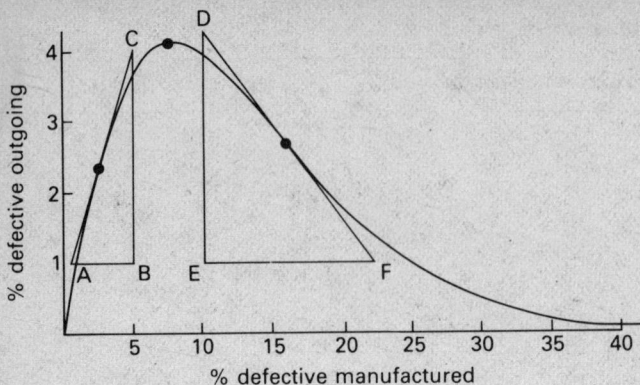

Fig. 9.5 Percentage defective manufactured and outgoing after inspection

Exercises

1. Table 9.5 gives, for each height, two weights between which 80% of women of that height were found to weigh.

Table 9.5 Heights and weights of women aged 16–64, 1980

Height group (cm)	150.1–155.0	155.1–160.0	160.1–165.0	165.1–170.0	170.1–175.0
Group midpoint (cm)	152.55	157.55	162.55	167.55	172.55
80% are between	45.0	48.5	52.0	52.5	57.0
and	70.5	73.0	76.0	78.0	81.0
(kg)					

Source: Adult Height and Weight Survey OPCS

   (a) Draw the graph in Fig. 9.1 and add to it the information in Table 9.5.

   (b) Record the height and weight of three women that you know, mark them on your graph and comment. (Perhaps not to them!)

2. Table 9.6 shows the cost per item for various batch sizes of a mass-produced item. Clearly the more that are made in one batch, the cheaper each one is to produce, since any initial set-up costs are shared among more items.

   (a) Draw the graph and comment.

   (b) Estimate the cost per item if a batch of 750 is made.

   (c) Estimate the cost per item if a batch of 5000 is made.

   (d) Estimate the gradients for batch sizes of 200, 500 and 1000.

Table 9.6 Costs for various batch sizes

Batch size	Cost per item (£)
100	8.00
200	5.50
300	4.67
400	4.25
500	4.00
1000	3.50
2000	3.25

3. Table 9.7 shows various combinations of the amount advanced by a building society as a standard repayment mortgage and the term of years of the loan for a monthly repayment of £100.00.

Table 9.7 Loan and repayment term for a monthly repayment of £100.00

Repayment term (years)	Loan (£)
10	7 716.05
15	9 699.32
20	10 989.01
25	11 834.32
30	12 376.24

(a) Draw the graph and comment.
(b) With a monthly repayment of £100 over 18 years estimate the amount advanced.

4. The depth of water over a harbour bar at various times of days shown in Table 9.8.
(a) Draw the graph of this data and comment.
(b) Estimate the times of high tide and low tide.

Table 9.8 Depth of water at different times

Time (hrs)	07.00	08.00	09.00	10.00	11.00	12.00	13.00	14.00	15.00	16.00	17.00
Depth (m)	11.39	12.88	13.10	12.01	9.89	7.31	4.96	3.48	3.25	4.35	6.61

   (c) During which times is the tide rising (positive gradient) and falling (negative gradient)?

   (d) Estimate the greatest rate of decrease in depth and say when this occurs.

Unit 10
Simple algebra

Whenever letters or symbols are used to represent numbers, or sets of numbers then the resulting expressions are called *algebraic* expressions. For example,

$$C = \frac{5}{9}(F-32),$$

where C is degrees Celsius and F is degrees Fahrenheit, or

$$A = \pi r^2,$$

where A is the area of a circle with radius r, are both examples of algebraic expressions. Algebra enables answers to be found to problems by simple operations with letters rather than by repeated arithmetic with numbers.

An algebraic expression can be considered to be a set of letters and numbers combined by the arithmetic operators $+$, $-$, $\times$, $\div$. For example, in the manufacture of a certain product the profit in pounds, obtained by making and selling x products may be represented by

$$2x - 50.$$

This is an algebraic expression with two terms, $2x$ and 50. The *terms* of an algebraic expression are those parts of it which are connected by $+$ or $-$ signs. Further definitions necessary for the understanding of algebra are
1. *Variables*: letters used to represent different numbers
   e.g. in $2x - 50$ there is just one variable, x.
2. *Coefficient*: a number placed before, and thus multiplying a letter or group of letters
   e.g. in $2x - 50$ the coefficient of x is 2.
3. *Constant term*: a term with no variables
   e.g. in $2x - 50$, a constant term is -50.

Summarizing,

$$\text{coefficient} \rightarrow 2x - 50 \leftarrow \text{constant}$$
$$\text{of } x \qquad \uparrow$$
$$\text{variable}$$

It is particularly important that the meaning of algebraic expressions is clearly understood. Some examples of algebraic expressions, together with their meanings, are given in Table 10.1. Note that in algebra the multiplication sign, $\times$, is usually omitted to save confusion with x.

Table 10.1 Algebraic terminology

Algebraic expression	Interpretation
x	$1(x)$
$2x$	$2(x)$ or $x + x$
ax	$a(x)$
$\frac{1}{2}x$	$\frac{1}{2}(x)$ or $x/2$
$-x$	$-1(x)$ or $-1x$
$2ax$	$(2a)x$ or $ax + ax$
x^2	$x(x)$
$2x^2$	$2x(x)$ or $x^2 + x^2$
$x^{\frac{1}{2}}$	$\sqrt{x}$
$(2x)^2$	$(2x)(2x)$ or $(2^2)x^2$
$2ax^2$	$2a(x^2)$ or $ax^2 + ax^2$
$2x + 3y$	$2(x) + 3(y)$

Substitution is the replacing of letters in an algebraic expression by given values to obtain a numerical value for that expression. The algebraic expression that determines the temperature in degrees Celsius from the temperature in degrees Fahrenheit was stated at the beginning of this unit. What is the equivalent temperature in degrees Celsius when the temperature is 59 °F? Replace F in the expression

$$C = \frac{5}{9}(F - 32)$$

with the value 59. Then

$$C = \frac{5}{9}(59 - 32)$$
$$= \frac{5}{9}(27) = 15.$$

Therefore 59 °F is equivalent to 15 °C. In a similar way, the equivalent temperature to 70 °F is

$$C = \frac{5}{9}(70-32)$$

$$= \frac{5}{9}(38) = 21.1,$$

that is, 21.1 °C.

Example

The area of a trapezium is given by the algebraic expression

$$\frac{1}{2}(x+y)h$$

where x and y are the lengths of its two parallel sides and h is the perpendicular distance between these two sides. Find the area of the trapezium given in Fig. 10.1.

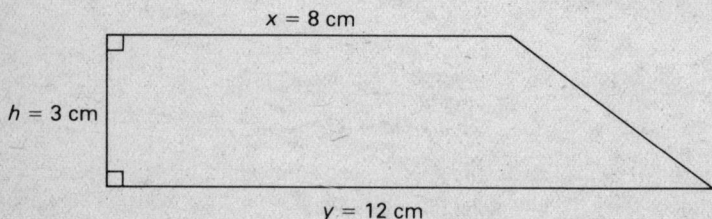

$x = 8$ cm

$h = 3$ cm

$y = 12$ cm

Fig. 10.1

Solution
In this example $x = 8$, $y = 12$, $h = 3$, so the area of the trapezium is

$$\frac{1}{2}(8+12)3 = \frac{1}{2}(20)3 = (10)3 = 30 \text{ cm}^2.$$

Example

If $x = 3$, $y = 2$, and $z = -1$, find the values of
(a) $2xy + z$
(b) $3x^2 + 2yz$
(c) $4x - \frac{1}{2}y + (2z)^2$

Solution
(a) $2xy + z = 2(3)(2) + (-1)$
$\qquad\qquad = 12 - 1 = 11$

(b) $3x^2 + 2yz = 3(3^2) + 2(2)(-1)$
$\qquad\qquad = 27 - 4 = 23$

(c) $4x - y + (2z)^2 = 4(3) - \frac{1}{2}(2) + (2(-1))^2$
$\qquad\qquad\qquad = 12 - 1 + (-2)^2$
$\qquad\qquad\qquad = 12 - 1 + 4 = 15$

Do not forget that $(-2)^2 = (-2)(-2) = 4$ as multiplication involving two negative numbers gives a positive answer.

Suppose we wish to evaluate

$5x - 3x$

when $x = 7$. It could be done by substituting $x = 7$ into this expression

$5(7) - 3(7) = 35 - 21 = 14.$

However, it is more straightforward to simplify the original expression to get

$5x - 3x = (5-3)x$
$\qquad\quad = 2x$

and then substitute $x = 7$ into this expression,

$2(7) = 14.$

The aim of simplification is to convert an algebraic expression into one that is shorter and easier to handle. The terms $5x$ and $-3x$ are called *like terms* because the only difference is their coefficients. For example,

$3y, 5y,$ and $-6y$

are like terms. Only like terms can be added or subtracted as a single term, the process being called 'collecting like terms together'. For example,

$8a + 6a - 3a + a = 12a.$

However, $3x$ and $2y$ are not like terms and so $3x + 2y$ cannot be simplified in this way. However, the order in which the letters are written in like terms is not important, for example,

$4xyz - 2yzx + 7zyx = 9xyz$

as xyz, yzx and zyx are also like terms. A useful tip when dealing with more complicated terms is always to write the letters in alphabetical order.

When multiplying or dividing algebraic expressions containing like letters, the basic rules of powers as described in Unit 4 apply. Therefore

$$(9x^2yz)(2xy^3z) \div 3xyz$$
$$= 18x^3y^4z^2 \div 3xyz$$
$$= \frac{18x^3y^4z^2}{3xyz}$$
$$= 6x^2y^3z$$

Example

Simplify
(a) $4x^2y - 3x^2y + x^2y$
(b) $4ab(-2bc)$
(c) $6mn^2 \div 2m^2n$

Solution
(a) $4x^2y - 3x^2y + x^2y$
     $= (4-3+1) \, x^2y$
     $= 2x^2y$
(b) $4ab(-2bc)$
     $= -8ab^2c$
(c) $6mn^2 \div 2m^2n$
     $= \frac{6mn^2}{2m^2n} = \frac{3n}{m}$

Brackets, or parentheses, are often used in algebraic expressions as they are when performing numerical calculations. They may be removed by multiplying the term outside the bracket by each of the terms inside the bracket. For example

$$3(x+y) = 3x + 3y.$$

In particular, care must be taken with the sign of each term when removing brackets; a negative sign outside a bracket changes all the signs inside the bracket. For example,

$$-3(x-y+z) = -3x + 3y - 3z.$$

Example

Simplify
(a) $5x^2 - 2x(4-3x) - 3x$
(b) $3xy - y(2x+4)$

Solution
(a) $5x^2 - 2x(4-3x) - 3x$
$\quad = 5x^2 - 8x + 6x^2 - 3x$
$\quad = 11x^2 - 11x$
which could be written $11x(x-1)$.
Do not be overambitious when simplifying algebraic expressions involving brackets. First remove the brackets and then collect like terms together. Do not attempt to carry out both processes in one step.
(b) $3xy - y(2x+4)$
$\quad = 3xy - 2xy - 4y$
$\quad = xy - 4y$
$\quad = (x-4)y$

The product of two brackets is obtained by multiplying each term in the first bracket by each term in the second. For example,

$$(a+b)(x+y) = ax + bx + ay + by$$

Example

Expand
(a) $(x-1)(x+2)$
(b) $(a-b)(a+b)$
(c) $(x+y+z)^2$

Solution
(a) $(x-1)(x+2)$
$\quad = x^2 - x + 2x - 2$
$\quad = x^2 + x - 2$
(b) $(a-b)(a+b)$
$\quad = a^2 - ba + ab - b^2$
$\quad = a^2 - b^2$
(c) $(x+y+z)^2$
$\quad = (x+y+z)(x+y+z)$
$\quad = x^2 + yx + zx + yx + y^2 + yz + zx + zy + z^2$
$\quad = x^2 + 2xy + 2xz + y^2 + 2yz + z^2$

Algebra is a useful tool for translating verbal statements into mathematical form. For example, if one kilo of apples costs 48p, an algebraic expression for the cost of x kilos is (in pence) $48x$.

Similarly, if one kilo of carrots cost 35 pence and a cauliflower costs

18 pence each then the total cost of p kilos of carrots and q cauliflowers is (in pence)

$35p + 18q$.

When forming an algebraic expression from a description it is a useful idea to substitute numerical values into the expression to check that it is correct.

Example

Under normal driving conditions, an approximation for the minimum braking distance (in metres) of a car is obtained by dividing the square of the speed (in km/hr) by 200.
(a) Express a relationship between the minimum braking distance, d metres, and the speed, v km/hr.
(b) Find the minimum braking distances from speeds of 40 km/hr and 80 km/hr.

Solution
(a) $d = \dfrac{1}{200} v^2$

(b) when $v = 40$, $d = \dfrac{1}{200} (40)40 = 8$ metres,

$v = 80$, $d = \dfrac{1}{200} (80)80 = 32$ metres.

Example

Mr Davies keeps a very careful record of his motoring expenses. The fixed costs of possessing his car (which includes road tax and insurance) are £180 per year. His variable costs (which includes petrol, oil and maintenance) are 8 pence per mile. There is public transport from his home to all the places he normally travels to by car, the fare amounting to 15 pence per mile.
(a) Using the symbol x to represent the number of miles travelled by Mr Davies in a year, give an expression for his annual travelling costs
   (i) if he only travels by car,
   (ii) if he sells his car and travels by public transport.
(b) Decide which method of transport costs less if he travels
   (i) 2000 miles,
   (ii) 5000 miles.

Solution

(a) (i) If he travels by car,
  Cost (in £) = 180 + 0.08x
  Be careful with units.
  (ii) Cost (in £) = 0.15x
  if public transport is used.

(b) (i) If x = 2000
  Cost using car = 180 + 0.08(2000) = £340
  Cost using public transport = 0.15(5000) = £750
  If the distance travelled is 2000 miles, public transport is cheaper.
  (ii) If x = 5000
  Cost using car = 180 + 0.08(5000) = £580
  Cost using public transport = 0.15(5000) = £750
  If the distance travelled is 5000 miles, car travel is cheaper.

(No account has been taken of the money obtained from the sale of his car.)

Exercises

1. The volume of a cylinder is given by
$$v = \pi r^2 h,$$
where r is the radius of the cylinder and h is the height of the cylinder. Calculate the volume of a cylindrical tank with height 3 metres and radius $\frac{1}{2}$ metre.

2. The relationship between °F and °C is given by
$$C = \frac{5}{9}(F - 32)$$
If $F = 45°$ determine the equivalent temperature in °C.

3. If $x = 5$, $y = -2$, find values of
   (a) $5x^2 + 2xy - y^2$
   (b) $3x^2y$
   (c) $5x + 4y - 8$

4. Simplify (a) $8x - 5x + 3x$
   (b) $2x^2y - 5x^2y + yx^2$
   (c) $5x + 8y$

5. Simplify (a) $4xy(2x^2y)$
   (b) $8ab^3 \div 2a^3b$
   (c) $5p(8q) \div 4p/q$

6. One gallon of petrol costs £1.80 and one can of oil costs 60 pence. Write down an expression for the cost of x gallons of petrol and y cans of oil.
Determine this cost when $x = 6$ and $y = 2$.

7. Expand (a) $(x+y)^2$
           (b) $5x - 2(x-2) - 5$
           (c) $(a+b-ab)(a-b)$

8. A small manufacturer makes electrical components. He has fixed costs of £800 per month (including rent and rates) and variable costs of 60 pence per component. At the end of each month a wholesaler buys as many components as the company can produce, paying £1 per component. If the manufacturer makes x components in a month, write algebraic expressions for
(a) the monthly costs,
(b) the monthly revenue.
Determine the monthly costs and monthly revenue if he manufactures 1000, 2000 and 3000 components. Comment on your answer.

9. Simplify $2(x^2-x+3) - (x^2+2x-4)$.

10. A rocket accelerates upwards such that h, its height in metres, after t seconds is given approximately by the formula
$h = 12t^2$.
Calculate its height after 5 seconds and after 20 seconds. What is the average velocity during this time period? Obtain an algebraic expression that determines the average velocity, between time a seconds and b seconds.

Unit 11
Linear equations

The monthly repayments needed to repay a mortgage over 25 years are shown in Table 11.1. This has already been discussed in Units 4, 6 and 9, and it is clear from Fig. 9.3 that the relationship between the amount of loan and the monthly repayment is linear, i.e. the graph is a straight line. Before discussing this further, it is necessary to introduce some symbols: let L be the size of the loan, measured in thousands of pounds and R the monthly repayment in pounds, so that we have, from Table 11.1, if $L = 5$, $R = 42.25$.

Table 11.1 Calendar monthly repayments at 8.925% p.a.

Amount of loan (£000's)	1	5	10	20
Monthly repayment (£)	8.45	42.25	84.50	169.00

To find the repayment necessary for a loan of £16,500 the graph could be used, as in Unit 9, but this will only give an approximate answer. A better approach would be to find an equation that will allow R to be calculated from L. Now when $L = 1$, $R = 8.45$ and when L is increased by a multiple of 10 to 10, R is increased by a multiple of 10 to 84.50. Also when L doubles from 10 to 20, so does R from 84.50 to 169.00. These results are the consequences of the linear relationship between L and R, and it is possible to deduce that if $L = 2$, then R will be $2(8.45) = 16.9$, while if $L = 3$, $R = 3(8.45) = 25.35$. The algebraic relationship between L and R, or the linear equation giving R in terms of L, is

$$R = 8.45(L)$$
or $R = 8.45L$

This equation can be used to give accurate values for the

monthly repayments. A loan of £16,500 ($L = 16.5$) will need a monthly repayment of $R = 8.45(16.5) = 139.425$, i.e. £139.43, while a loan of £37,950 ($L = 37.95$) will need a monthly repayment of $R = 8.45(37.95) = 320.678$ i.e. £320.68. (This explains the accurate results stated in Unit 9.)

To manufacture a large batch of electronic components there are two costs incurred, a fixed set up cost of £500.00 and a variable cost due to labour and materials of £3.00 for each component. Hence if a batch of 100 components is made the total cost is $500 + 3(100) = £800.00$. Let the number of components in a batch be x and let the total cost be C. By calculating C for various values of x the results in Table 11.2 were obtained, and the graph of these results is in Fig. 11.1. This graph shows that there is a linear relationship between C and x.

Table 11.2 Costs for various batch sizes

Batch size, x	10	100	200	300
Cost, C (£)	530	800	1100	1400

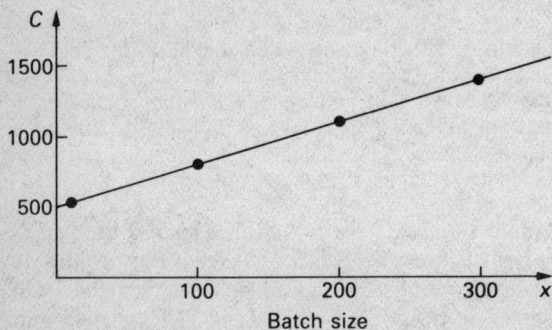

Fig. 11.1 Costs for various batch sizes

The linear equation connecting C and x is

$C = 500 + 3x$.

You should use this equation to check the results in Table 11.2; for example, if $x = 200$,

$C = 500 + 3(200)$
i.e. $C = 500 + 600$
∴ $C = 1100$.

If the total cost is to be £1,010, what should the batch size be? This is equivalent to finding x when $C = 1010$, or solving the equation

$500 + 3x = 1010$.

The graph can be used to give an approximate answer by reading off the value of x that corresponds to $C = 1010$, but a more accurate method is to use the rules of algebra (Unit 10). An equation remains true as long as whatever is done to one side is also done to the other. Subtracting 500 from both sides produces:

$3x = 510$.

Dividing both sides by 3 produces:

$$x = \frac{510}{3} = 170.$$

Thus a cost of £1010 is incurred if a batch of 170 is produced.

Example

The two variables x and y are related by the following linear equation:

$y = 15 - 2x$

(a) Find y if $x = 4$
(b) Find x if $y = 5$
(c) Find x if $y = 0$

Solution

(a) If $x = 4$, then $y = 15 - 2(4)$
$\qquad\qquad\qquad = 15 - 8$
$\qquad\qquad\qquad = 7$
$\qquad\qquad \therefore\ y = 7$ when $x = 4$.

(b) If $y = 5$, then $5 = 15 - 2x$
$\quad$ Add $2x$ to both sides:$\quad 5 + 2x = 15$
$\quad$ Subtract 5 from both sides: $2x = 10$
$\quad$ Divide both sides by 2:$\qquad x = 5$
$\qquad\qquad \therefore\ x = 5$ when $y = 5$.

(c) If $y = 0$, then $0 = 15 - 2x$
$\qquad\qquad \therefore\ 2x = 15$
$\qquad\qquad \therefore\quad x = \frac{15}{2}$
$\qquad\qquad \therefore\quad x = 7.5$ when $y = 0$.

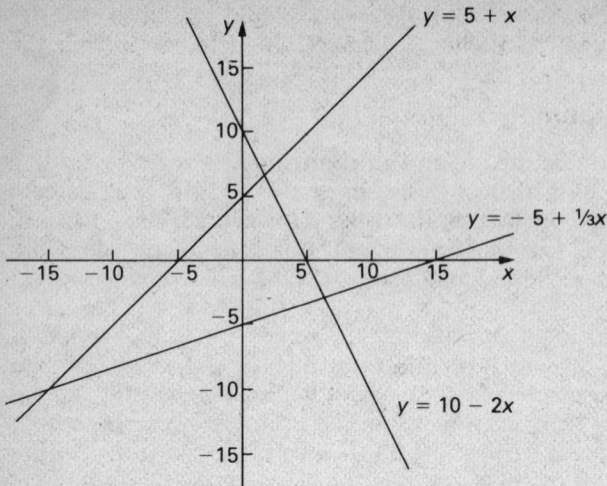

Fig. 11.2

Both of the previous linear equations have been of the form $y = a + bx$ where a and b are constants. For example, the equation $y = 15 - 2x$ has $a = 15$ and $b = -2$. The equation $y = a + bx$ is typical of linear equations which predict y from x, and Fig. 11.2 shows the graphs of three such equations, each of which is discussed in detail.

1. $y = 5 + x$. Here $a = 5$ and $b = 1$. If $x = 0$, $y = 5$, the value of a. If $x = 10$, $y = 15$, and so y increases by 10 when x increases by 10. Hence the gradient is $\dfrac{10}{10} = 1 = b$.

2. $y = -5 + \dfrac{1}{3}x$. Here $a = -5$ and $b = \dfrac{1}{3}$. If $x = 0$, $y = -5$, the value of a. If $x = 10$, $y = -5 + \dfrac{10}{3} = -5 + 3\dfrac{1}{3} = -1\dfrac{2}{3}$ and so y increases by $3\dfrac{1}{3}$ when x increases by 10. Hence the gradient is $\dfrac{3\frac{1}{3}}{10} = \dfrac{10}{3} \div 10 = \dfrac{10}{3}\left(\dfrac{1}{10}\right) = \dfrac{1}{3} = b$.

3. $y = 10 - 2x$. Here $a = 10$ and $b = -2$. If $x = 0$, $y = 10$, the value of a. If $x = 10$, $y = 10 - 20 = -10$, and so y decreases by 20 when x increases by 10. Hence the gradient is $-\dfrac{20}{10} = -2 = b$.

These three examples illustrate the general result that for the graph of the linear equation $y = a + bx$,

a is the *intercept* on the y axis (value of y when $x = 0$), and
b is the *gradient*.

Example

The relationship between x and y is linear. When $x = 0$, $y = 3$, and when $x = 4$, $y = 19$. Find the equation relating y to x.

Solution

Knowing two points that the straight line goes through, the graph could be drawn, since it is a straight line. From this graph the intercept (a) and gradient (b) could be measured, and so the equation $y = a + bx$ determined. In this example, however, both a and b can be found without a graph. Since $y = 3$ when $x = 0$, the intercept is 3 i.e. $a = 3$. Since y increases by $19 - 3 = 16$ when x increases by 4, the gradient is $\frac{16}{4} = 4$. Thus $b = 4$ and the required equation is:

$$y = 3 + 4x.$$

A gardener is planning to grow a mixture of gooseberry and blackcurrant bushes on a plot of land of area 300 m^2. Each gooseberry bush requires 3 m^2 while each blackcurrant bush requires 5 m^2. Let G and B be the numbers of gooseberry and backcurrant bushes respectively. Then if all the 300 m^2 is to be used, the following linear equation must hold:

$3G + 5B = 300$.
If $G = 0$, then $5B = 300$ or $B = 60$
If $B = 0$, then $3G = 300$ or $G = 100$
If $B = 30$, then $3G + 5(30) = 300$
 i.e. $3G + 150 = 300$
Subtract 150 from both sides: $3G = 150$
Divide both sides by 3: $G = 50$.

These results show three possible alternatives. Grow just blackcurrant (60 bushes), or grow just gooseberries (100 bushes) or grow a mixture of 30 blackcurrant and 50 gooseberries. Figure 11.3 shows the linear relationship between B and G and shows that there are many alternatives open to the gardener. This example will be discussed further in Unit 12.

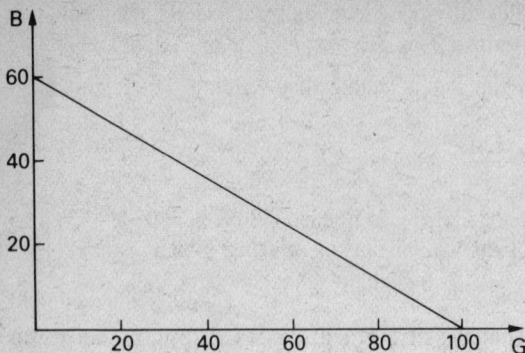

Fig. 11.3 Numbers of blackcurrant and gooseberry bushes in 300 m^2

Example

If $5x + 7y = 114$,

(a) find x if $y = 2$

(b) find y if $x = 6$

(c) find an equation for y in the form $y = a + bx$

(d) find the gradient, i.e. the rate of change of y with respect to x.

Solution

(a) If $y = 2$, then $5x + 7(2) = 114$

$$5x + 14 = 114$$

Subtract 14 from both sides: $5x = 100$

Divide both sides by 5: $x = \dfrac{100}{5} = 20$

$$\therefore \text{ If } y = 2, x = 20$$

(b) If $x = 6$, then $5(6) + 7y = 114$

$$30 + 7y = 114$$

Subtract 30 from both sides: $7y = 84$

Divide both sides by 7: $y = \dfrac{84}{7} = 12$

$$\therefore \text{ If } x = 6, y = 12$$

(c) $5x + 7y = 114$

Subtract $5x$ from both sides: $7y = 114 - 5x$

Divide both sides by 7: $y = \dfrac{114}{7} - \dfrac{5}{7}x$

Thus either $y = \dfrac{114}{7} - \dfrac{5}{7}x$

or $y = 16.29 - 0.71x$.　　(Accurate to 2 d.p.)

(d) The gradient is b, and so the answer is $-\frac{5}{7}$ or -0.71.

Example
Solve the following equation
$$7(x-3) = 10 + 4(x-1)$$

Solution
First multiply out the brackets, giving
$$7x - 21 = 10 + 4x - 4$$

Add 21 to both sides:
$$7x = 10 + 4x - 4 + 21$$
$$\text{i.e. } 7x = 4x + 10 - 4 + 21$$
$$7x = 4x + 27$$

Subtract $4x$ from both sides:
$$3x = 27$$

Divide both sides by 3:
$$x = 9.$$

Exercises

1. A firm has determined that the costs of ordering are given by a fixed cost independent of the number of orders placed, and a variable cost proportional to the number of orders placed. When 10 are ordered, the total cost is £105, and when 20 are ordered the total cost is £205. If x represents the number ordered and C the total cost (£), show that the linear equation $C = 5 + 10x$ fits the given data and hence
   (a) Identify the fixed costs
   (b) Evaluate the cost of ordering 5
   (c) Find how many could be ordered for a cost of £255.

2. If $y = 3x + 7$
   (a) find y if $x = 2$
   (b) find x if $y = 7$.

3. If $y = 4x - 5$, which one of the following is true?
   (a) $x = 4y - 5$
   (b) $x = 4y + 5$

(c) $x = \dfrac{1}{4}y + 5$

(d) $x = \dfrac{1}{4}y + \dfrac{5}{4}$

4. The two variables p and q are linearly related. When $q = 0$, $p = 4$, and when $q = 3$, $p = 19$. Find an equation of the form $p = a + bq$ which fits this information.

5. The variables C and t are linearly related. When $t = 1$, $C = 9$ and when $t = 3$, $C = 35$. Find an equation of the form $C = a + bt$ which fits this information. (You may need to draw a graph.)

6. If $2x + 3y = 60$,
   (a) find x if $y = 10$
   (b) find y if $x = 9$
   (c) find an equation for y in the form $y = a + bx$
   (d) find the gradient, i.e. the rate of change of y with respect to x.

7. Solve $(x+2)3 - 4 = 9x - 10$.

8. Solve $2(x-1) = 3(x-2)$.

9. Solve $\dfrac{2}{x-3} = \dfrac{3}{x-2}$

10. Solve $\dfrac{x-3}{5} = \dfrac{17-x}{2}$.

Unit 12
Simultaneous equations

Consider again the gardener who wants to plant an area of 300 m^2 with gooseberry and blackcurrant bushes. This situation was first discussed in Unit 11, where a linear equation connecting the two variables G and B, the numbers of gooseberry and blackcurrant bushes, was derived. This equation arose because each gooseberry bush needs 3 m^2 of land whilst each blackcurrant bush needs 5 m^2, and all the available 300 m^2 of land is to be used. This produces the linear equation

$$3G + 5B = 300 \qquad\qquad\qquad\qquad [1]$$

If ever this equation is referred to again, it can be called equation [1], and a similar notation will be used for future equations in this unit. The graph of equation [1] is a straight line passing through the points $(G=0, B=60)$, $(G=100, B=0)$ and $(G=50, B=30)$ (see Fig. 11.3). Clearly there are many different pairs of values for G and B that satisfy equation [1].

Now besides being restricted to 300 m^2, the gardener has just £120.00 to spend on fruit bushes, with each gooseberry bush costing £1.00 and each blackcurrant bush costing £3.00. This cost information means that there is a second linear equation connecting G and B which is

$$G + 3B = 120 \qquad\qquad\qquad\qquad [2]$$

The graph of equation [2] is also a straight line and again there are many different pairs of values for G and B that satisfy it. The important question for the gardener is whether he can find a pair of values for G and B such that both equations are true, i.e. all his land and cash are exactly used up. Stated mathematically, this problem is to *solve the two simultaneous equations*:

$$3G + 5B = 300 \qquad\qquad\qquad\qquad [1]$$
$$G + 3B = 120 \qquad\qquad\qquad\qquad [2]$$

They are called simultaneous because any solution, i.e. a pair of values for G and B, must satisfy them both simultaneously. Such simultaneous equations can be solved either graphically or algebraically, and both methods of solution are given here.

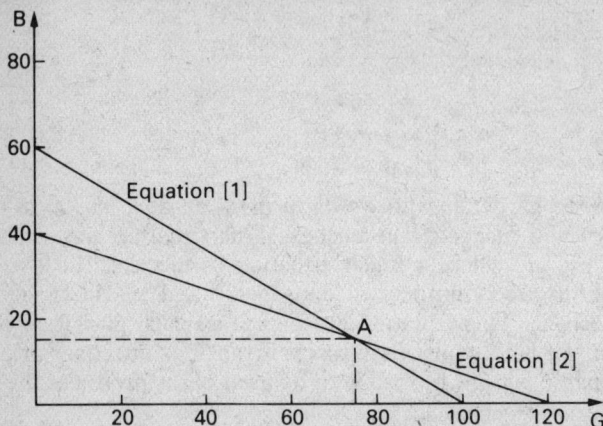

Fig. 12.1 Numbers of gooseberry and blackcurrant bushes

Figure 12.1 shows both equation [1] and equation [2] drawn on the same set of axes. To do this the scale on both axes has to be such that both lines will fit on the graph, and this may require a little trial and error. It can be seen from these graphs that at only one point, A, are both equations satisfied by the same values for G and B. The values for G and B at this intersection point can be read off and they provide approximate answers to the original problem of solving a pair of simultaneous equations. Hence the solution is $G = 75$, $B = 15$, i.e. buy 75 gooseberry bushes and 15 blackcurrant bushes. You can verify that with these values for G and B both equation [1] and equation [2] are satisfied exactly.

This graphical method is not only approximate, but can be time-consuming, and the following algebraic method is preferred, being accurate and quick once the basic technique is mastered. Since each equation has two variables in it, it is not possible to solve either individually, but if a new equation could be constructed which had only one variable, then this could be solved using the methods of Unit 10 and 11. This is the basic approach, and the first aim is to create a pair of equations from the original two so that in these new equations the coefficients of B are the same (or the coefficients of G

are the same). Refer to equations [1] and [2]. To make the G terms the same, leave [1] as it is and multiply both sides of equation [2] by 3. To make the B terms the same it would mean multiplying both sides of [1] by 3 and both sides of [2] by 5, which works, but is not as convenient as the first method. Here is the formal solution for solving these two equations, which are given again for convenience:

$$3G + 5B = 300 \qquad [1]$$
$$G + 3B = 120 \qquad [2]$$

Multiply [2] by 3: $\qquad 3G + 9B = 360 \qquad [3]$
Rewrite [1] $\qquad\qquad 3G + 5B = 300 \qquad [1]$
Subtract [1] from [3]: $\quad 0 + 4B = 60$
Divide both sides by 4: $\qquad B = \dfrac{60}{4} = 15$

Substitute $B = 15$ into [1]:

$$3G + 5(15) = 300$$
$$3G + 75 = 300$$

Subtract 75 from both sides:

$$3G = 225$$

Divide both sides by 3:

$$G = \frac{225}{3} = 75$$

The solution is $B = 15$, $G = 75$, which is the same as that obtained graphically. The above method is typical of the solution of all simultaneous linear equations and the following are some comments on it:

1. After inspection of [1] and [2] it was decided to eliminate G rather than B, although elimination of B would produce the same correct solution.
2. Subtracting [1] from [3] is the key step. Since by definition both sides of any equation are equal, this is just the same as subtracting the same thing from both sides of equation [3].
3. Having found a value for B, G can be found by substituting this value for B into either of the two original equations. Had [2] been used, this would produce

$\therefore \ G + 3(15) = 120$
$\therefore \ G + 45 = 120$
$\therefore \ G = 75$, as before.

111

4. Having obtained a solution, it can be checked by verifying that both of the original equations are satisfied. Substituting $G = 75$ and $B = 15$ into [1] gives:

$$3(75) + 5(15) = 225 + 75 = 300$$

Substituting $G = 75$ and $B = 15$ in [2] gives:

$$75 + 3(15) = 75 + 45 = 120$$

Example

At a party held in a public house, the first round bought cost £8.54 and was for four pints of bitter and seven pints of lager. The second round was for six pints of bitter and five pints of lager and cost £8.30. (Two people found the lager too gassy!) What are the prices per pint of bitter and lager?

Solution

Let x and y be the price, in pence, of a pint of bitter and lager respectively. Each round then allows an equation to be made, using pence as the standard unit, not pounds.

Round 1	$4x + 7y = 854$	[1]
Round 2	$6x + 5y = 830$	[2]

We need to solve these simultaneous equations to find x and y. (An algebraic solution is presented here, but you may like to try it using a graph). The easiest variable to eliminate is x, and although this could be done by multiplying [1] by 6 and [2] by 4, a neater way is to multiply [1] by 3 and [2] by 2.

Multiply [1] by 3: $12x + 21y = 2562$ [3]
Multiply [2] by 2: $12x + 10y = 1660$ [4]
Subtract [4] from [3]:

$$11y = 902$$

Divide both sides by 11:
$$y = \frac{902}{11} = 82$$

Substitute $y = 82$ into [1]:

$$4x + 7(82) = 854$$
$$4x + 574 = 854$$

Subtract 574 from both sides:

$4x = 280$

Divide both sides by 4:

$$x = \frac{280}{4} = 70$$

Thus bitter is 70p/pint, lager is 82p/pint. This is a cheaper pub than the 'Fox and Hounds' discussed in Unit 1! Check by substitution into equations [1] and [2].

Into [1]: $4(70) + 7(82) = 280 + 574 = 854 \checkmark$
Into [2]: $6(70) + 5(82) = 420 + 410 = 830 \checkmark$

Example

Solve the following simultaneous equations by first eliminating y.

$$2x - 4y = -12 \qquad\qquad [1]$$
$$3x + 2y = 22 \qquad\qquad [2]$$

Solution
Multiply [2] by 2: $6x + 4y = 44$ [3]
Rewrite [1]: $2x - 4y = -12$ [1]
Add [3] to [1]: $8x = 32$
Divide both sides by 8:

$x = 4$

Substitute $x = 4$ into [1]:

$2(4) - 4y = -12$
$8 - 4y = -12$

Subtract 8 from both sides:

$-4y = -12 - 8$
$-4y = -20$

Divide both sides by -4:

$$y = \frac{-20}{-4} = +5$$

Solution: $x = 4$, $y = 5$

If any of the manipulations with negative signs in this solution are not clear, read Unit 1.

Example

Solve the following simultaneous equations by first eliminating x.

$$2x - 4y = -12 \qquad [1]$$
$$3x + 2y = 22 \qquad [2]$$

Solution

Multiply [1] by 3: $6x - 12y = -36 \qquad [3]$
Multiply [2] by 2: $6x + 4y = 44 \qquad [4]$
Subtract [4] from [3]

$$-12y - 4y = -36 - 44$$
$$\therefore -16y = -80$$

Divide both sides by -16:

$$y = \frac{-80}{-16} = +5$$

Substitute $y = +5$ into [1]:

$$2x - 4(5) = -12$$
$$2x - 20 = -12$$

Add 20 to both sides:

$$2x = -12 + 20$$
$$2x = 8$$

Divide both sides by 2:

$$x = 4$$

Solution: As before, $x = 4$ and $y = 5$.

Example

Solve for x and y:

$$3x + 8 = 2y \qquad [1]$$
$$5y = 1 - 2x \qquad [2]$$

Solution

These two equations are not in the same form as those previously solved, and so the first step is to rewrite them.

$$3x + 8 = 2y \qquad [1$$

Subtract 8 from both sides: $3x = 2y - 8$
Subtract $2y$ from both sides: $3x - 2y = -8 \qquad [3$

Equation [3] is now in the form we are used to. Now to put [2] into the same form.

$$5y = 1 - 2x \qquad \text{[2]}$$

Add $2x$ to both sides: $\qquad 2x + 5y = 1 \qquad \text{[4]}$

Rewrite [3]: $\qquad\qquad\quad 3x - 2y = -8 \qquad \text{[3]}$

Equations [3] and [4] can now be solved as before.

Multiply [4] by 3: $\qquad\quad 6x + 15y = 3 \qquad \text{[5]}$

Multiply [3] by 2: $\qquad\quad 6x - 4y = -16 \qquad \text{[6]}$

Subtract [6] from [5]: $\qquad 15y - (-4y) = 3 - (-16)$

$$15y + 4y = 3 + 16$$
$$19y = 19$$
$$\therefore \quad y = 1$$

Substitute $y = 1$ into [1]: $\quad 3x + 8 = 2(1)$

$$3x + 8 = 2$$

Subtract 8 from both sides: $\; 3x = 2 - 8$

$$3x = -6$$

Divide both sides by 3: $\qquad x = \dfrac{-6}{3} = -2$

Solution: $x = -2, y = 1$.

Exercises

1. When some items are offered for sale the quantity demanded (q) by customers will decrease if the price (p) increases. However, as the price increases the quantity supplied by the retailer will increase. For a particular item the demand and supply equations are:
   Demand: $q + 2p = 1000$
   Supply: $q - 2p = -200$
   (a) Graph both these on the same axes and find the values of q and p (in pence) which make supply equal demand.
   (b) Solve these equations algebraically to confirm your solution from (a). (Eliminate p first.)

2. Solve the following simultaneous equations by first eliminating B, and confirm that you get the same solution as that given in the text.
   $3G + 5B = 300$
   $G + 3B = 120$

3. The sum of two numbers is 20 and their difference is 4. Write this as two simultaneous equations and hence find the two numbers.

4. Solve for x and y: $\qquad 3x - 4y = 5$
   $$2x + 3y = 26$$

5. Solve for p and q:

$$2p - 3q = 5$$
$$5p + 2q = 22$$

6. Solve for a and b:

$$-3a + 4b = 5$$
$$2a - 5b = -15$$

7. Solve for x and y:

$$7x - 2y = 1$$
$$-2y + 7y = 64$$

8. Solve for x and y:

$$4x = 3y + 10$$
$$-x = -2y + 10$$

9. Solve for x and y:

$$x + 6 = y$$
$$3y - 16 = 4x$$

10. Solve for x and y:

$$4x = 7 + 3y$$
$$-6y + 8x = 15$$

Unit 13
Quadratic equations

When a cricket ball is hit over a flat pitch, the relationship between the height of the ball above the ground, H metres, and the time the ball has been in the air, t sec, can be shown to be (approximately)

$$H = 5t(5-t).$$

It can be seen from this equation that when $t = 0$, $H = 0$, i.e. initially the ball is at ground level, while if $t = 5$, $H = 0$ again, showing that after a flight lasting 5 secs, the ball is again at ground level. The results of evaluating H for values of t ranging from 0 to 5 are shown in Table 13.1, and the graph of these results is shown in Fig. 13.1. This curve is a typical example of a *parabola* or parabolic curve.

Table 13.1 Height and time for a cricket ball

Time (t sec)	0	1	2	3	4	5
Height (H m)	0	20	30	30	20	0

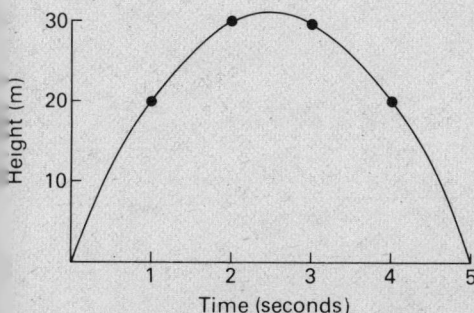

Fig. 13.1 Height and time for a cricket ball

Consider the problem of trying to find when the ball is at a height of 20 m. From either Table 13.1 or Figure 13.1 it can be seen that the ball is at a height of 20 m twice, first when $t = 1$ and then when $t = 4$. When is the ball at a height of 25 m? This is not clear from the table, but the graph does allow two approximate answers to be obtained, the first after just less than 1.5 sec and then after slightly more than 3.5 sec. In this unit more accurate methods of obtaining these results are discussed and to illustrate these methods the problem of finding t when $H = 20$ will be solved, although the answer of $t = 1$ or $t = 4$ is already known. (The problem when $H = 25$ will be solved later as an example.)

The original equation with $H = 20$ is

$$20 = 5t(5-t)$$

and after multiplying out the brackets, this becomes

$$20 = 25t - 5t^2.$$

Add $5t^2$ to both sides:

$$5t^2 + 20 = 25t$$

Subtract $25t$ from both sides:

$$5t^2 - 25t + 20 = 0$$

Divide both sides by 5:

$$t^2 - 5t + 4 = 0 \qquad [1]$$

This equation is a typical example of a quadratic equation, i.e. an equation where the most complicated term is the variable squared.

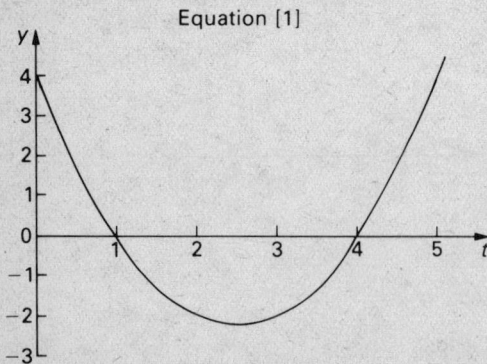

Fig. 13.2

Solving equation [1] should produce the two solutions already known, $t = 1$ and $t = 4$, and three methods of solving this and similar equations are now discussed.

The graphical method requires the graph of

$$y = t^2 - 5t + 4$$

to be drawn, and then this graph is used to read off the values of t for which $y = 0$. By evaluating y for various values of t, the graph in Fig. 13.2 is produced, and this clearly shows that when $y = 0$, $t = 1$ or $t = 4$, which confirms the earlier solutions. Any quadratic equation like equation [1] will allow a graph to be drawn, and the solutions will be where the curve crosses the horizontal axis. There are three possibilities:

1. Two distinct solutions, as in Fig. 13.2 and Fig. 13.3(a).
2. One single solution, as in Fig. 13.3(b).
3. No solution, as in Fig. 13.3(c).

This graphical method, although instructive, can only produce approximate solutions, and one of the following two algebraic methods should be used.

Equation [1] can be solved by rewriting it in terms of its two factors. Since

$$t^2 - 5t + 4 = (t-1)(t-4),$$

equation [1] is

$$(t-1)(t-4) = 0.$$

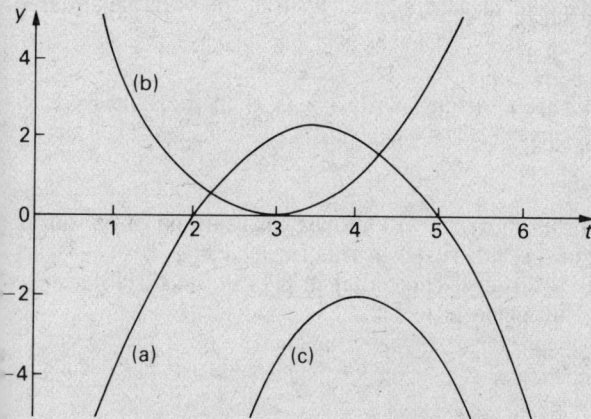

Fig. 13.3   Three possible types of solution

Whenever the product of two terms is zero, then either the first or the second (or both) must be zero.

Hence either $(t - 1) = 0$ or $(t - 4) = 0$
either $t = 1$ or $t = 4$.

Not all quadratic equations can be factorized, and even when they can it is not always easy to obtain the two factors. It is really a matter of trial and error, testing each possibility suggested. For example, consider factorizing and hence solving

$$x^2 + 3x - 10 = 0. \qquad\qquad [2]$$

The only way x^2 can be split into its factors is by multiplying x and x, while -10 could be either $(-10,1)$, $(10,-1)$, $(-5,2)$ or $(5,-2)$. This means that there are four possibilities for the two factors of [2]:

$(x-10)(x+1) = x^2 + x \; - 10x - 10 = x^2 - 9x - 10$ (NO)
$(x+10)(x-1) = x^2 - x \; + 10x - 10 = x^2 + 9x - 10$ (NO)
$(x-5) \; (x+2) = x^2 + 2x - \; 5x - 10 = x^2 - 3x - 10$ (NO)
$(x+5) \; (x-2) = x^2 - 2x + \; 5x - 10 = x^2 + 3x - 10$ (YES)

Thus the factors of [2] are

$(x+5)(x-2) = 0$
either $x + 5 = \;\;\; 0$ or $x - 2 = 0$
either $\qquad x = -5$ or $\qquad x = 2$

The third and final method for solving quadratics is to apply the following formula, which can be proved to give the solutions, if there are any, of any quadratic equation. When the quadratic equation is written in the form

$$ax^2 + bx + c = 0,$$

where a, b and c are constants with $a \neq 0$, then the solution is

$$x = \frac{-b \pm \sqrt{b^2 - 4ac}}{2a}$$

(The plus or minus sign, $\pm$, will allow two solutions to be found.)

When equation [1] is written in this form, $a = 1$, $b = -5$ and $c = 4$, and so the solution is (note that in [1] t is the variable, not x, but this has no important effect):

$$t = \frac{-(-5) \pm \sqrt{(-5)^2 - 4(1)(4)}}{2(1)}$$

i.e. $t = \dfrac{+5 \pm \sqrt{25 - 16}}{2}$

i.e. $t = \dfrac{+5 \pm \sqrt{9}}{2}$

$t = \dfrac{+5 \pm 3}{2}$

Thus either $t = \dfrac{+5+3}{2}$ or $t = \dfrac{+5-3}{2}$

either $t = \dfrac{8}{2}$ or $t = \dfrac{2}{2}$

either $t = 4$ or $t = 1$.

Example

A small rectangular lawn is to be laid with a total area of 40 m^2, and with the length 3 m more than the width. Find the width.

Solution
Let the width be x, and so the length must be $x + 3$ and the area must be $x(x+3)$. Thus

$x(x+3) = 40$

i.e. $x^2 + 3x = 40$

i.e. $x^2 + 3x - 40 = 0$ [3]

Equation [3] is a quadratic. Trying to find the factors produces $x^2 + 3x - 40 = (x+8)(x-5)$.

Hence $(x+8)(x-5) = 0$

$\therefore$ either $x + 8 = 0$ or $x - 5 = 0$

$\therefore$ either $x = -8$ or $x = 5$

There is only one solution, $x = 5$, since $x = -8$ is clearly not practicable. If finding the factors is too difficult, then the formula can always be used, and in this case $a = 1, b = 3, c = -40$, giving

$x = \dfrac{-3 \pm \sqrt{(3)^2 - 4(1)(-40)}}{2(1)}$

i.e. $x = \dfrac{-3 \pm \sqrt{9 + 160}}{2}$

i.e. $x = \dfrac{-3 \pm \sqrt{169}}{2}$

$x = \dfrac{-3 \pm 13}{2}$

$\therefore$ either $x = \dfrac{-3 + 13}{2}$ or $x = \dfrac{-3 - 13}{2}$

121

either $x = \dfrac{10}{2}$ or $x = \dfrac{-16}{2}$

either $x = 5$ or $x = -8$

Again the solution is $x = 5$.

Example

When a car travelling at v km/hour brakes to a stop on a dry road, the approximate distance taken to stop, D metres, which includes the thinking time of the driver, is given by:

$$D = \frac{v}{5} + \frac{v^2}{170}$$

This equation gives a better approximation than the one discussed in Unit 10. If in an accident a car travelled 60 metres before stopping, estimate the speed that the car must have been doing before the accident.

Solution

We need to solve

$$60 = \frac{v}{5} + \frac{v^2}{170}$$

Multiply both sides by 170:

$$10,200 = 34v + v^2$$

Subtract 10,200 from both sides:

$$0 = 34v + v^2 - 10200$$

Rearrange:

$$v^2 + 34v - 10,200 = 0$$

Solving this equation by finding the factors is clearly at best difficult, and not worth the time. In this case, using the formula is a more direct method, with $a = 1$, $b = 34$ and $c = -10,200$.

$$v = \frac{-34 \pm \sqrt{(34)^2 - 4(1)(-10,200)}}{2(1)}$$

$$v = \frac{-34 \pm \sqrt{1156 + 40,800}}{2}$$

$$v = \frac{-34 \pm \sqrt{41,956}}{2}$$

$$v = \frac{-34 \pm 204.83}{2}$$

So either $v = \dfrac{-34 - 204.83}{2}$ or $v = \dfrac{-34 + 204.83}{2}$

either $v = \dfrac{-238.83}{2}$ or $v = \dfrac{170.83}{2}$

either $v = -119.41$ or $v = 85.41$

Hence $v = 85.4$ km/hour. The negative answer is clearly not practical, and since one decimal place is enough accuracy in a final answer for speed, two decimal places were used in the solution.

Example
Solve

$$x^2 - 6x + 9 = 0$$

Solution
This equation can be factorized, giving

$$x^2 - 6x + 9 = (x-3)(x-3),$$

and so $(x-3)(x-3) = 0$,
    either $x - 3 = 0$ or $x - 3 = 0$
    i.e. either $x = 3$ or $x = 3$.

This is an example of a quadratic equation with a single solution (see Fig. 13.3 (b)), although strictly it should be called a repeated solution.

Example
After how many seconds will the cricket ball discussed earlier be at a height of 25 m?

Solution
Using the equation at the start of this unit with $H = 25$ gives:

$$25 = 5t(5-t)$$
i.e. $25 = 25t - 5t^2$

Divide both sides by 5:

$$5 = 5t - t^2$$

Add t^2 to both sides:

$$t^2 + 5 = 5t$$

Subtract $5t$ from both sides:

$$t^2 - 5t + 5 = 0$$

This quadratic cannot be factorized simply, so the formula should be used with $a = 1$, $b = -5$, and $c = 5$, giving

$$t = \frac{-(-5) \pm \sqrt{(-5)^2 - 4(1)(5)}}{2(1)}$$

i.e. $t = \dfrac{+5 \pm \sqrt{25 - 20}}{2}$

i.e. $t = \dfrac{+5 \pm \sqrt{5}}{2}$

$t = \dfrac{5 \pm 2.236}{2}$

either $t = \dfrac{7.236}{2}$ or $t = \dfrac{2.764}{2}$

either $t = 3.618$ $\qquad t = 1.382$

(Compare these results with the approximate ones obtained by a graphical method earlier.)

Example

Solve $\quad x^2 + x + 1 = 0$

Solution

This will not factorize, so use the equation with $a = 1$, $b = 1$ and $c = 1$, giving

$$x = \frac{-1 \pm \sqrt{(1)^2 - 4(1)(1)}}{2(1)}$$

i.e. $x = \dfrac{-1 \pm \sqrt{1 - 4}}{2}$

i.e. $x = \dfrac{-1 \pm \sqrt{-3}}{2}$

Now the square root of a negative number does not exist, so there are no values of x that satisfy this equation. This situation can often occur, and can be understood best by considering a graphical attempt to find a solution in which the curve does not cross or touch the horizontal axis, as in Fig. 13.3(c).

The three possibilities for solving a quadratic

$$ax^2 + bx + c = 0$$

can now be described by:

1. Two distinct solutions, as in Fig. 13.3(a), when $b^2 - 4ac$ is positive.
2. One repeated solution, as in Fig. 13.3(b), when $b^2 - 4ac$ is zero.
3. No solution, as in Fig. 13.3(c), when $b^2 - 4ac$ is negative.

Example

Solve $x^2 - 3x = 0$

Solution

By factoring, $x^2 - 3x = x(x-3)$, and so

$$x(x-3) = 0$$

∴ either $x = 0$ or $x - 3 = 0$
∴ either $x = 0$ or $x = 3$

Do not ignore the solution $x = 0$ in problems like this.

Exercises

1. Draw the graph of $y = 2x^2 - 11x + 12$ for x between 0 and 6 and hence estimate the values of x that satisfy the quadratic equation $2x^2 - 11x + 12 = 0$

2. By factorizing $x^2 - 3x + 2$ solve the equation $x^2 - 3x + 2 = 0$

3. By using an appropriate formula solve $x^2 - 3x - 3 = 0$

4. Obtain an accurate solution to question 1.

5. Use an alternative method to solve question 2, so confirming the previous solution.

6. Use factorization to solve the following equations
   (a) $x^2 + 3x + 2 = 0$
   (b) $p^2 - 3p = 0$
   (c) $6x^2 - 13x + 6 = 0$
   (d) $3n^2 + 11n - 4 = 0$
   (e) $-2t^2 - t + 21 = 0$

7. Use an appropriate formula to solve, where possible, the following equations
   (a) $x^2 + x - 1 = 0$
   (b) $t^2 + 3t + 5 = 0$
   (c) $5p^2 + 5p - 5 = 0$
   (d) $0.73x^2 + 1.03x - 0.52 = 0$
   (e) $-2n^2 + 3n + 3 = 0$

8. Simplify and then solve, where possible, the following equations.
   (a) $x(x-2) = 1$
   (b) $2t(t+3) = t^2 + 4$
   (c) $\sqrt{x} = x - 10$
   (d) $(n+2)n = (n-2)(3-n)$
   (e) $\dfrac{3}{n} = n + 2$

Part four

Answers to exercises

Unit 1

1. (a) $7 + 5(2) = 7 + 10 = 17$
   (b) $3 + 7(4) = 3 + 28 = 31$
   (c) $5(3) + 6 = 15 + 6 = 21$
2. $35 - (-50) = 35 + 50 = 85$ metres
3. (a) $((-3) + 7) \div 2 = 4 \div 2 = 2$
   (b) $((-2)-(-4)3 = (-2+4)3 = (2)3 = 6$
   (c) $((-5)+(-3)) \div (-4) = (-5-3) \div (-4) = (-8) \div (-4)$
   $= \dfrac{-8}{-4} = \dfrac{8}{4} = 2$
4. (a) $44,800$
   (b) $45,000$
5. (a) $350 + 5(25) = 350 + 125 = 475$
   (b) $80(4) + 10 = 320 + 10 = 330$
   (c) $24(5) + 18(7) = 120 + 126 = 246$
6. $51,000$
7. (a) $(-40) + (-5)16 = -40 - 80 = -120$
   (b) $((-25)+(-365)) \div (-13) = (-25-365) \div (-13)$
   $= (-390) \div (-13)$
   $= \dfrac{-390}{-13} = \dfrac{390}{13} = 30$
8. $29,000 - (-36,000) = 29,000 + 36,000 = 65,000$ feet
9. $36 - (-52) = 36 + 52 = £88$
10. $83(36) \div (6+4(3)) = 83(36) \div (6+12)$
    $= 83(36) \div 18$
    $= \dfrac{83(36)}{18} = 166$
11. $645 - (124+43) = 645 - 167 = £478$
12. (a) Cost $= 323 + 4(62) + 2(83)$
    (b) $323 + 248 + 166 = £737$
    (c) $£740$

Unit 2

1. $109.2 - 103.0 = 6.2$
2. $123\frac{3}{4} \div 5\frac{1}{2} = \dfrac{495}{4} \div \dfrac{11}{2} = \dfrac{495}{4} \times \dfrac{2}{11} = \dfrac{45}{2} = 22\frac{1}{2}$
3. (a) $\dfrac{3}{4} + \dfrac{5}{8} = \dfrac{6}{8} + \dfrac{5}{8} = \dfrac{6+5}{8} = \dfrac{11}{8} = 1\frac{3}{8}$

(b) $\dfrac{5}{9} - \dfrac{7}{18} = \dfrac{10}{18} - \dfrac{7}{18} = \dfrac{10-7}{18} = \dfrac{3}{18} = \dfrac{1}{6}$

(c) $\dfrac{4}{7}\left(\dfrac{3}{8}\right) = \dfrac{4(3)}{7(8)} = \dfrac{3}{14}$

4. Total time $= 6\frac{3}{4} + 7\frac{1}{2} + 8\frac{5}{6} + 7\frac{1}{4} + 5\frac{2}{3}$

$= (6+7+8+7+5) + (\frac{3}{4}+\frac{1}{2}+\frac{5}{6}+\frac{1}{4}+\frac{2}{3})$

$= 33 + (\frac{9}{12}+\frac{6}{12}+\frac{10}{12}+\frac{3}{12}+\frac{8}{12})$

$= 33 + \dfrac{9+6+10+3+8}{12}$

$= 33 + \dfrac{36}{12}$

$= 33 + 3 = 36$ hours

$36(3.22) = £115.92$

5. (a) $9.421 - 5.3264 = 4.0946$

(b) $5.71 \div 21.5 = 0.266$ correct to 3 decimal places

(c) $4.29(7.31-4.216) = 4.29(3.094) = 13.27326$

6. $C = 2\pi r = 2\left(\dfrac{22}{7}\right)10\frac{1}{2} = 2\left(\dfrac{22}{7}\right)\dfrac{21}{2} = \dfrac{2(22)(21)}{7(2)} = (22)3 = 66$

Using calculator value of π, $C = 65.97$

7. Alternative method costs $= 24(17.19) = £412.56$

Increased cost $= 412.56 - 384.95 = £27.61$

8. Total cost $= 1.41 + 2(1.82) + 2.23 + 2.65 + 2.80 + 2.95 + 3.45$

$= 1.41 + 3.64 + 2.23 + 2.65 + 2.80 + 2.95 + 3.45$

$= £19.06$

9. (a) A litre bottle contains 1000 ml.

The number of measures $= \dfrac{1000}{25} = 40$

(b) Revenue per bottle $= 40 \times 0.55 = £22.00$

Contribution to profit $=$ Revenue $-$ Cost $= 22 - 7.20$
$= £14.80$

(c) Fraction $= \dfrac{7.20}{22} = \dfrac{36}{110} = \dfrac{18}{55}$

10. Number of hours asleep $= \dfrac{1}{4}(24) = 6$

Number of hours at work $= \dfrac{3}{8}(24) = 9$

Number of hours travelling $= \dfrac{1}{12}(24) = 2$

Number of free hours $= 24 - (6+9+2) = 24 - 17 = 7$

Fraction of free time $= \dfrac{7}{24}$

(Answer could have been obtained from $1 - (\frac{1}{4}+\frac{3}{8}+\frac{1}{12})$)

11. 9.6(42.7) = 409.92 miles

12. $250 \div 3\frac{1}{8} = 250 \div \frac{25}{8} = \frac{250(8)}{25} = 10(8) = 80$

Unit 3

1. (a) 1 : 2   (b) 15 : 2   (c) 4 : 1
2. 3750 : 125 is equivalent to 30 : 1
3. 200 : 120 is equivalent to 5 : 3
4. $7 + 3 = 10$, so $\frac{7}{10}$ yellow and $\frac{3}{10}$ blue

   Amount of yellow paint $= \frac{7}{10}(50) = 35$ litres

   Amount of blue paint $= \frac{3}{10}(50) = 15$ litres

5. Amount of nitrogen $= 2 \times$ amount of potash
   $$= 2(5) = 10 \text{ g}$$
   Amount of phosphorus $=$ amount of potash
   $$= 5 \text{ g}$$

6. $8 + 3 + 6 = 17$

   Alliance received $\frac{8}{17}(34,000) = 16,000$ votes

   Conservatives received $\frac{3}{17}(34,000) = 6000$ votes

   Labour received $\frac{6}{17}(34,000) = 12,000$ votes

7. Actual distance $= 2.5(2) = 5$ km between Exford and Wyemouth distance on map $17.5(\frac{1}{2}) = 8.75$ cm

8. (a) $\frac{1}{4}(100) = 25\%$
   (b) $0.1(100) = 10\%$
   (c) $0.167(100) = 16.7\%$

9. (a) $\frac{54}{100} = 0.54$

   (b) $\frac{8.3}{100} = 0.083$

   (c) $\frac{40}{100} = 0.4$

10. (a) $\frac{50}{100}(84) = 42$

(b) $\dfrac{35}{100}(16) = 5.6$

(c) $\dfrac{120}{100}(45) = 54$

(d) $\dfrac{3}{100}(700) = 21$

(e) $\dfrac{50}{100}\left(\dfrac{70}{100}(180)\right) = \dfrac{50}{100}(126) = 63$

11. If 18 are absent, $300 - 18 = 282$ are present

Percentage present $= \dfrac{282}{300}(100) = 94\%$

12. $\dfrac{2}{100}(750) = £15$

13. Reduction in price $= \dfrac{20}{100}(250) = 50$

Price of television $= 250 - 50 = £200$

14. Meal $= £15$

VAT $= \dfrac{15}{100}(15) = £2.25$

So meal + VAT $= 15 + 2.25 = £17.25$

Service charge $= \dfrac{10}{100}(17.25) = £1.72$

Total cost of meal $= 17.25 + 1.72 = £18.97$

15. $\dfrac{12}{100}(4000) = 480$

So value of car at end of first year $= 4000 - 480 = £3520$

$\dfrac{12}{100}(3520) = 422.40$

So value of car at end of second year $= 3520 - 422.40 = £3097.60$

Percentage of initial cost $= \dfrac{3097.60}{4000}(100) = 77.44\%$

Unit 4

1. (a) 512   (b) 1   (c) 16
   (d) 10,000   (e) 7
2. (a) $4(4^2) = 4^1 4^2 = 4^3 = 64$
   (b) $6^3 \div 6^5 = \dfrac{6^3}{6^5} = \dfrac{1}{6^2} = \dfrac{1}{36}$

Polish up your maths

(c) $7^2(7^{-3}) = \frac{7^2}{7^3} = \frac{1}{7}$

(d) $(-3)^{-2} = \frac{1}{(-3)^2} = \frac{1}{9}$

(e) $3^{\frac{1}{2}}(3^{2\frac{1}{2}}) = 3^3 = 27$

3. $5^{10} = 9,765,625$

4. (a) 8   (b) 6.325   (c) Does not exist

(d) $12.6^{\frac{1}{2}} = \sqrt{12.6} = 3.55$ (e) $7.98^{\frac{1}{3}} = \sqrt[3]{7.98} = 1.998$

5. $10,000(1.08)^3 = 10,000(1.259712) = £12,597.12$

6. (a) $0.01(0.01) = 0.0001$

(b) $\sqrt{0.01} = 0.1$

(c) $0.01^{-1} = \frac{1}{0.01} = 100$

7. $p = 750, r = 9.5, n = 1.5$

$$P\left(1 + \frac{r}{100}\right)^n = 750\left(1 + \frac{9.5}{100}\right)^{1.5} = 750(1.095)^{1.5}$$
$$= 750(1.1458326) = £859.37$$

8. $n = 25, r = 8.4, P = 50,000$

Annual repayments $= \frac{50,000(8.4)}{100}\left(\frac{1.084^{25}}{1.084^{25}-1}\right)$

$= \frac{50,000(8.4)}{100}\left(\frac{7.511593112}{6.511593112}\right)$

$= 500(8.4)(1.153572249)$

$= £4845.00$

Monthly repayments $= \frac{4845}{12} = £403.75$

9. (a) 54,230   (b) 72.9   (c) 0.0861

10. $0.475 - 0.09324 = 0.38176 = 3.8176(10^{-1})$

Unit 5

1. (a) Mean temperature $= \frac{3.8+3.5+4.6+ \ldots +5.7}{12} = \frac{118.3}{12}$

$= 9.86\ °C$

(b) Mean rainfall $= \frac{144+57+59+ \ldots +81}{12} = \frac{899}{12} = 74.92\ mm$

(c) Mean sunshine $= \frac{2.18+2.14+2.09+ \ldots +1.76}{12} = \frac{51.37}{12}$

$= 4.28\ hrs$

2. Size 7 is the shoe most frequently sold
    7 is the mode.

3. Mean score $= \dfrac{9+32+23+\ldots+26}{18} = 493 = 27.4$

4. In ascending order
   35 36 36 37 38 38 39 39 41 41 42 42 43 44 47
   Median = 41

5. Mean price $= \dfrac{384+396+354+\ldots+360}{18} = \dfrac{3024}{8} = 378$

6. In ascending order
   621, 681, 752, 826, 873, 942, 1036, 1092
   Median = (826+873) = 849.5

7. (a) Mean $= \dfrac{1.77+1.69+1.72+\ldots+1.76}{8} = \dfrac{21.12}{12} = £1.76$

   (b) In ascending order
   1.69, 1.72, 1.73, 1.76, 1.77, 1.77, 1.77, 1.77, 1.77, 1.78, 1.79, 1.80
   Median = £1.77

   (c) Most frequently occurring price = £1.77
   As a large proportion of garages have the price £1.77, the mode is the most suitable measure of average.

8. Mean number of children per houshold
   $= \dfrac{2+1+0+\ldots+2}{30} = \dfrac{66}{30} = 2.2$
   Most frequent number in the table = 2. Mode = 2

Unit 6

1. (a) From the middle column, the monthly repayments are

      £72.80

   (b) For a £10,000 loan over 30 years the monthly repayments are £80.80. For £24.000 monthly repayments are

      2.4(80.80) = £193.92

   (c) Over 10 years, for a loan of £7,000, repayments are £90.72 and for a loan of £8,000, repayments are £103.68.
      As 7,500 is halfway between 7,000 and 8,000 then the

      monthly repayments $= \dfrac{90.72+103.68}{2} = £97.20.$

   (d) Over 10 years, £6,000 need repayments of £77.76, and over 15 years the monthly repayments are £61.86, a difference of 5 years and a difference of £15.90. As 12 years is 3 years below the lower repayment, an estimate of monthly repayments over 12 years is

$$61.86 + \frac{3}{5}(15.90) = £71.40$$

2. (a) £39   (b) £66

3.

Class	Tally	Frequency
10–19	11	2
20–29	1111	4
30–39	1111 11	7
40–49	1111 1111	10
50–59	1111 1111 1111	14
60–69	1111 1111	9
70–79	1111 111	8
80–89	1111	5
90–99	1	1

4. (a) Adding the five numbers
753.7 + 730.8 + 719.2 + 721.5 + 729.6
= 3654.8
So the total number of births during this five-year period
= 3,654,800

(b) 1981 $\dfrac{188.5}{730.8} \times 100 = 25.8\%$

1982 $\dfrac{185.9}{719.2} \times 100 = 25.8\%$

1983 $\dfrac{187.4}{721.5} \times 100 = 26.0\%$

1984 $\dfrac{191.0}{729.6} \times 100 = 26.2\%$

Unit 7

1. (a) August and July are the most popular and November is the least popular.
   (b) August is the most popular and November, December and January are the least popular.
   (c) Not necessarily, since this chart shows only percentages, and the total going abroad may be much higher, even though the percentage is lower.

2. (a) Highest is 18–24 year age group and lowest the 65 and over.

(b) Highest is again the 18–24 year age group but the lowest is the 55–64 year group.

(c) No! This chart shows that for those who drink, men drink twice as much as women, but the numbers of each sex who do drink would need to be known before such a statement could be confirmed.

3. (a) 55 and over.

(b) Yes.

(c) Males spend nearly twice as much time on drink related activities, while females watch much more TV.

(d) Yes.

Unit 8

1.

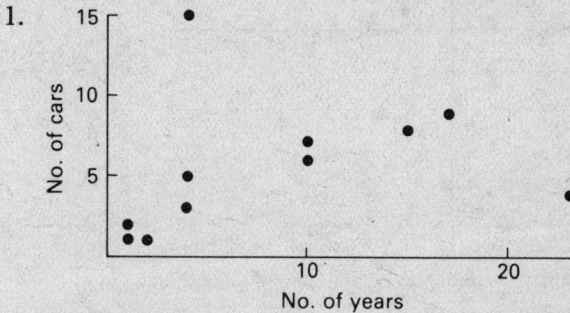

Fig. 1 Number of cars and driving experience

With the exception of D and G, there is a clear relationship or trend.

4.

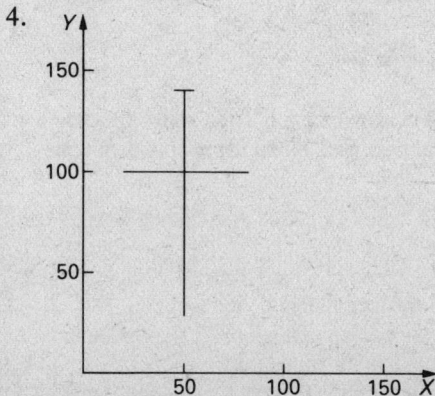

Fig. 2 Computer graphics

5. MOVE 20,20
   DRAW 20,120
   DRAW 70,120
   MOVE 20,70
   DRAW 50,70
6. (a) C   (b) A   (c) D   (d) F
7.

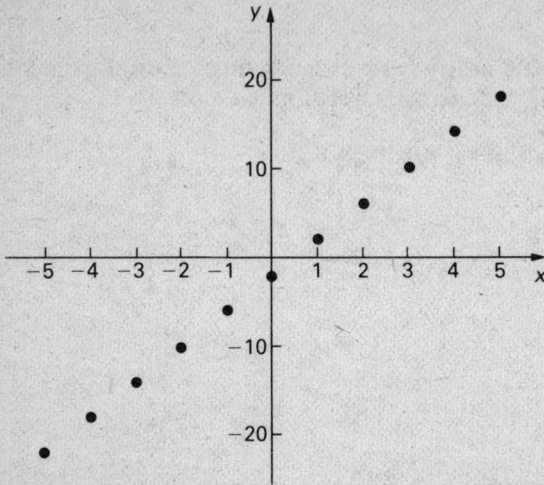

Fig. 3   Relationship $y = 4x - 2$

8. (a) T   (b) T   (c) F   (d) T   (e) T
9. MOVE 20,20
   DRAW 20,120
   MOVE 20,70
   DRAW 70,70
   MOVE 70,20
   DRAW 70,120
   This draws the letter H. To now draw E first write instructions
   for an E starting at 20,20, then add 80 to all x-coordinates.
   MOVE 80,20
   DRAW 80,120
   DRAW 130,120
   MOVE 80,70
   DRAW 110,70
   MOVE 80,20
   DRAW 130,20
   For the A, add 160 after starting at 20,20, etc.

Unit 9

1. (a)

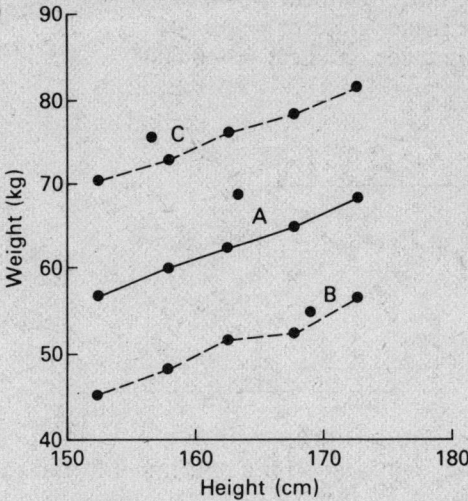

Fig. 4 Height and weight of women

(b) Person A is fairly typical, B is rather thin, while C is clearly overweight.

2. (a)

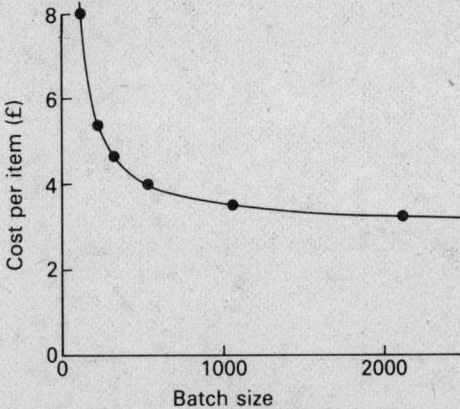

Fig. 5 Cost for various batch sizes

The cost per item is decreasing, but it should not drop below £3.00.

(b) £3.67
(c) £3.10
(d) When 200 are made, gradient = −0.013
When 500 are made, gradient = −0.002
When 1000 are made, gradient = −0.0005

3. (a)

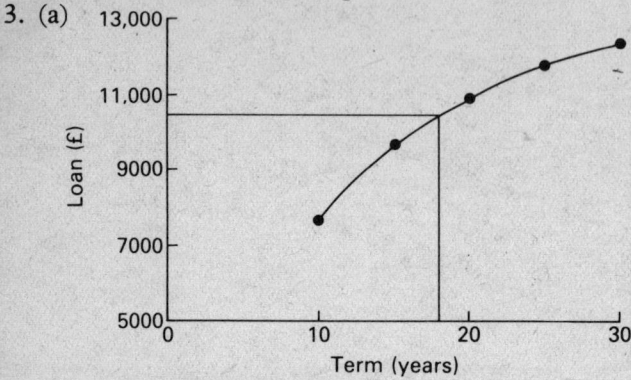

Fig. 6 Loan and repayments term for a repayment of £100.00

Nearly a straight line.
(b) £10,450 (approximately)

4. (a)

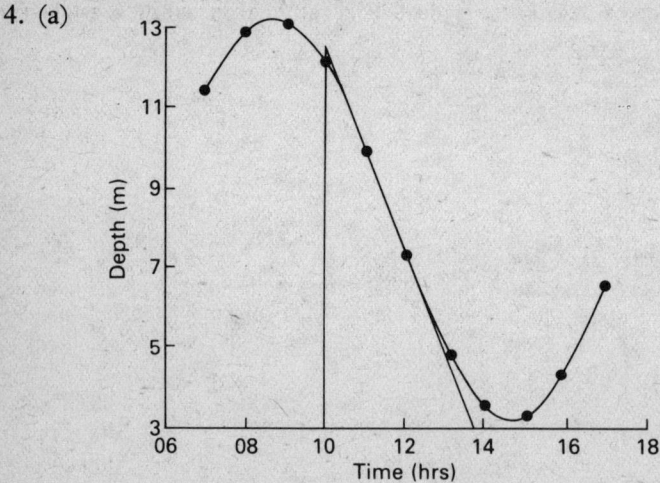

Fig. 7 Depth of water over time

(b) High tide is at 0840
Low tide is at 1440

138

(c) The water rises from 0700 to 0840 and from 1440 to 1700. The water falls from 0840 to 1440.

(d) The greatest rate of decrease is at 1140 hours and is
$$\frac{9.5}{3.7} = 2.6 \text{ m/hour (1 d.p.).}$$

Unit 10

1. $h = 3 \; r = \frac{1}{2}$
   $v = \pi \frac{1}{4} 3 = \frac{3}{4}\pi = 2.356 \text{ m}^3$

2. $C = \frac{5}{9}(45-32) = \frac{5}{9}(13) = 7.2 \,^\circ C$

3. (a) $5(5^2) + 2(5)(-2) - (-2)^2$
      $= 125 - 20 - 4 = 101$
   (b) $3(5^2)(-2) = -150$
   (c) $5(5) + 4(-2) - 8 = 25 - 8 - 8 = 9$

4. (a) $8x - 5x + 3x = (8-5+3)x = 6x$
   (b) $2x^2y - 5x^2y + yx^2 = (2-5+1)x^2y = -2x^2y$
   (c) $5x + 8y$ cannot be simplified

5. (a) $8x^3y^2$
   (b) $8ab^3 \div 2a^3b = \frac{8ab^3}{2a^3b} = 4\frac{b^2}{a^2}$
   (c) $5p(8q) \div 4p/q = 40pq\left(\frac{q}{4p}\right) = 10q^2$

6. Costs (£) = $1.8x + 0.6y$
   when $x = 6$, $y = 2$, Cost = $1.8(6) + 0.6(2) = 10.8 + 1.2 =$ £12.

7. (a) $(x+y)^2 = (x+y)(x+y) = x^2 + xy + yx + y^2 = x^2 + 2xy + y^2$
   (b) $5x - 2(x-2) - 5 = 5x - 2x + 4 - 5 = 3x - 1$
   (c) $(a+b-ab)(a-b) = a^2 - ab + ab - b^2 - a^2b + ab^2$
      $= a^2 - b^2 - a^2b + ab^2$

8. (a) Cost (£) = $800 + 0.6x$
   (b) Revenue (£) = x

x	1000	2000	3000
Costs	1400	2000	2600
Revenue	1000	2000	3000

   If x is more than 2000 then the manufacturer makes a profit.

9. $2(x^2-x+3)-(x^2+2x-4)$
   $= 2x^2 - 2x + 6 - x^2 - 2x + 4$
   $= x^2 - 4x + 10$

10. $h = 12t^2$
   when $t = 5$ $h = 300$ m
   when $t = 20$ $h = 4800$ m
   Difference = 4500 m in 15 seconds
   Average speed $= \dfrac{4500}{15} = 300$ m/sec
   Average speed $= \dfrac{12b^2 - 12a^2}{(b-a)}$ which can be simplified to
   $12(b+a)$.

Unit 11

1. If $C = 5 + 10x$, then
   when $x = 10$, $C = 5 + 10(10)$
           i.e. $C = 105$;
   when $x = 20$, $C = 5 + 10(20)$
           i.e. $C = 205$
   This confirms that the given equation fits the data.
   (a) Fixed costs are £5.00
   (b) If $x = 5$, then $C = 5+10(5) = 55$
       Thus ordering 5 costs £55
   (c) If $C = 255$, then
               $255 = 5 + 10x$
       Subtract 5 from both sides.
               $250 = 10x$
       Divide by 10
               $25 = x$
       Thus 25 can be ordered for £255.
2. (a) $y = 3(2) + 7 = 6 + 7 = 13$
       Hence if $x = 2$, $y = 13$.
   (b) $7 = 3x + 7$
       Subtract 7 from both sides.
               $0 = 3x$
       Divide by 3
               $0 = x$
       Hence if $y = 7$, $x = 0$.
3. (d)
4. $p = 4 + 5q$, since the intercept is 4 and the gradient is
   $\dfrac{(19-4)}{3} = +5$.

5. Having drawn a graph, a and b can be found to be -4 and $+13$ respectively, giving

$$C = -4 + 13t,$$

and the given values fit this equation.

6. (a) If $y = 10$, then
$$2x + 3(10) = 60$$
$$2x = 30$$
$$x = 15$$

   (b) If $x = 9$, then
$$2(9) + 3y = 60$$
$$3y = 42$$
$$y = 14$$

   (c) If $2x + 3y = 60$, then
$$3y = 60 - 2x$$
$$y = 20 - \frac{2}{3}x$$

   (d) The gradient is $-\frac{2}{3}$

7. $(x+2)3 - 4 = 9x - 10$
$$\therefore 3x + 6 - 4 = 9x - 10$$
$$3x + 2 = 9x - 10$$
Add 10 to both sides
$$3x + 12 = 9x$$
Subtract $3x$ from both sides
$$12 = 6x$$
Divide by 6
$$2 = x$$
i.e. $x = 2$.

8. $2(x-1) = 3(x-2)$
$$\therefore 2x - 2 = 3x - 6$$
$$2x + 4 = 3x$$
$$4 = x$$
i.e. $x = 4$.

9. $\dfrac{2}{x-3} = \dfrac{3}{x-2}$

Multiply both sides by $x - 3$
$$2 = \frac{3(x-3)}{x-2}$$
$$\therefore 2x - 4 = 3x - 9$$
$$-x = -5$$
i.e. $x = 5$

10. $\dfrac{x-3}{5} = \dfrac{17-x}{2}$

$\therefore\ 2(x-3) = 5(17-x)$

$\quad\ 2x - 6 = 85 - 5x$

$\qquad\quad 7x = 91$

$\qquad\qquad x = \dfrac{91}{7} = 13$

Unit 12

1. (a)

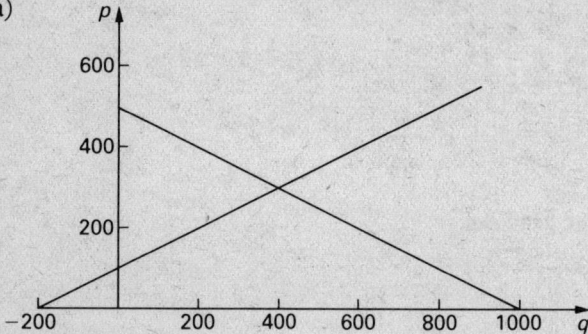

Fig. 8 Demand and supply

Supply equals demand when $q = 400$ and $p = 300$.

(b) $q + 2p = 1000$ [1]

$\quad\ q - 2p = -200$ [2]

Add [1] to [2]

$\quad\ 2q = 800$

$\therefore\quad q = 400$

Substitute $q = 400$ into [1]:

$\qquad\qquad 400 + 2p = 1000$

$\qquad\qquad\quad \therefore\ 2p = 600$

$\qquad\qquad\qquad\quad p = 300$

2. $3G + 5B = 300$ [1]

$\ \ G + 3B = 120$ [2]

Multiply [1] by 3 and [2] by 5:

$9G + 15B = 900$ [3]

$5G + 15B = 600$ [4]

Subtract [4] from [3]:

$\qquad 4G = 300$

$\therefore\quad G = 75$

142

Substitute $G = 75$ into [1]:
$225 + 5B = 300$
$\therefore 5B = 75$
$\therefore B = 15$

3. Let the two numbers be x and y. We now need to solve the following two simultaneous equations

$x + y = 20$ [1]
$x - y = 4$ [2]

Add [1] to [2]:
$2x = 24$
$\therefore x = 12$ and so $y = 8$

4. $3x - 4y = 5$ [1]
  $2x + 3y = 26$ [2]

Multiply [1] by 3 and [2] by 4:
$9x - 12y = 15$ [3]
$8x + 12y = 104$ [4]

Add [3] to [4]:
$17x = 119$
$x = 7$

Substitute $x = 7$ into [1]:
$21 - 4y = 5$
$-4y = -16$
$y = 4$

Solution: $x = 7, y = 4$

5. $2p - 3q = 5$ [1]
  $5p + 2q = 22$ [2]

Multiply [1] by 2 and [2] by 3:
$4p - 6q = 10$ [3]
$15p + 6q = 66$ [4]

Add [3] to [4]:
$19p = 76$
$p = 4$

Substitute $p = 4$ into [1]
$8 - 3q = 5$
$\therefore -3q = -3$
$q = 1$ Solution: $p = 4, q = 1$

6. Solution: $a = 5, b = 5$

7. Solution: $x = 3, y = 5$

8. Rewrite the equations as:
$4x - 3y = 10$
$-x + 2y = 10$
Solution: $x = 10, y = 10$.

9. Rewrite the equations as:
$$x - y = -6$$
$$-4y + 3y = 16$$
Solution: $x = 2, y = 8$

10. Rewrite the equations as
$$4x - 3y = 7 \qquad [1]$$
$$8x - 6y = 15 \qquad [2]$$
Multiply [1] by 2: $\qquad [3]$
$$8x - 6y = 14$$
Subtract [3] from [2]:
$$0 = 1!$$
Since it is not true, there is no solution, i.e. the two straight lines are parallel and do not cross.

Unit 13

1.

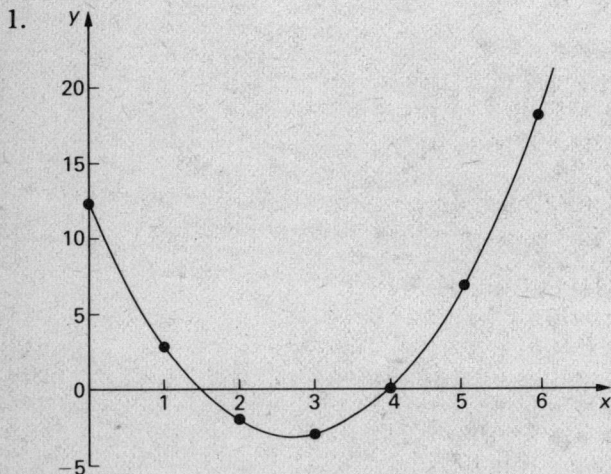

Fig. 9 Graph of $y = 2x^2 - 11x + 12$

From the graph, $x = 4$ or $x = 1.5$

2. $x^2 - 3x + 2 = (x-2)(x-1)$
Hence $(x-2)(x-1) = 0$
∴ either $x = 2$ or $x = 1$.

3. Using $x = \dfrac{-b \pm \sqrt{b^2 - 4ac}}{2}$ with

$a = 1, b = -3$ and $c = -3$ gives

$$x = \frac{+3 \pm \sqrt{9 - 4(1)(-3)}}{2(1)}$$

i.e. $x = \dfrac{3 \pm \sqrt{9 + 12}}{2}$

$\therefore \quad x = \dfrac{3 \pm \sqrt{21}}{2}$

$x = \dfrac{3 \pm 4.583}{2}$

$\therefore$ either $x = \dfrac{7.583}{2}$ or $x = \dfrac{-1.583}{2}$

Solution: Either $x = 3.79$ or $x = -0.79$ (2 d.p.)

4. $\qquad x = \dfrac{+11 \pm \sqrt{121 - 4(2)(12)}}{4}$

i.e. $x = \dfrac{+11 \pm \sqrt{25}}{4}$

either $x = \dfrac{11 + 5}{4}$ or $x = \dfrac{11 - 5}{4}$

Solution: Either $x = 4$ or $x = 1.5$

5. Using the formula,
$$x = \frac{+3 \pm \sqrt{9 - 8}}{2}$$

either $x = \dfrac{3 + 1}{2}$ or $x = \dfrac{3 - 1}{2}$

Solution: Either $x = 2$ or $x = 1$

6. (a) $x^2 + 3x + 2 = (x+2)(x+1)$
    either $x = -2$ or $x = 1$
   (b) $p^2 - 3p = p(p-3)$
    either $p = 0$ or $p = 3$
   (c) $6x^2 - 13x + 6 = (3x-2)(2x-3)$
    either $x = \frac{2}{3}$ or $x = \frac{3}{2}$
   (d) $3n^2 + 11n - 4 = (3n-1)(n+4)$
    either $n = \dfrac{1}{3}$ or $n = -4$
   (e) $-2t^2 - t + 21 = (-2t-7)(t-3)$
    either $t = -\dfrac{7}{2}$ or $t = 3$

7. (a) Either $x = 0.618$ or $x = -1.618$
   (b) Not possible to solve
   (c) Either $p = 0.618$ or $p = -1.618$ (see 7(a)!)
   (d) Either $x = 0.395$ or $x = -1.805$
   (e) Either $n = 2.186$ or $n = -0.686$

8. (a) $x^2 - 2x - 1 = 0$

$$x = \frac{+2 \pm \sqrt{4+4}}{2}$$

Either $x = 2.414$ or $x = -0.414$

(b) $2t^2 + 6t = t^2 + 4$

$t^2 + 6t - 4 = 0$

Either $t = 0.606$ or $t = -6.603$

(c) Square both sides

$$x = (x-10)^2$$

i.e. $x = x^2 - 20x + 100$

$x^2 - 21x + 100 = 0$

Either $x = 13.7$ or $x = 7.3$

(d) $n^2 + 2n = 3n - n^2 - 6 + 2n$

$2n^2 - 3n + 6 = 0$

No solution possible

(e) Multiply both sides by n

$3 = n^2 + 3n$

$n^2 + 3n - 3 = 0$

Either $n = 0.791$ or $n = 3.791$

Index